# Enhanced Recovery Methods for Heavy Oil and Tar Sands

# Enhanced Recovery Methods for Heavy Oil and Tar Sands

Editor

**Ajay Paralikar**

# Enhanced Recovery Methods for Heavy Oil and Tar Sands

Edited by **Ajay Paralikar**

Printed in 2017

ISBN: 978-1-68117-365-8

Library of Congress Control Number: 2015936529

© 2016 by
SCITUS Academics LLC,
616, Corporate Way, Suite 2, 4766,
Valley Cottage, NY 10989

www.scitusacademics.com

# Contents

# Preface

In Enhanced Recovery Methods for Heavy Oil and Tar Sands, Speight provides the current methods of recovery for heavy oil and tar sand bitumen technology, broken down by thermal and non-thermal methods. An engineer, graduate student or professional working with heavy oil, upcoming and current, will greatly benefit from this much-needed text. Recent oil price fluctuations continue to stress the need for more efficient recovery of heavy oil and tar sand bitumen resources. With conventional production steadily declining, advances in enhanced recovery will be required so that oil production can be extended and reservoirs last longer. A practical guide on heavy-oil related recovery methods is essential for all involved in heavy oil production. To feed this demand, James Speight, a well-respected scientist and author, provides a must-read for all scientists, engineers and technologists that are involved in production enhancement.

**Editor**

# Chapter 1

# Thermal Hydraulic Analysis Using GIS on Application of HTR to Thermal Recovery of Heavy Oil Reservoirs

Yangping Zhou, Fu Li, Zhiwei Zhou, and Yuanle Ma

Institute of Nuclear and New Energy Technology, Tsinghua University, Beijing 100084, China

## ABSTRACT

At present, large water demand and carbon dioxide ($CO_2$) emissions have emerged as challenges of steam injection for oil thermal recovery. This paper proposed a strategy of superheated steam injection by the high-temperature gas-cooled reactor (HTR) for thermal recovery of heavy oil, which has less demand of water and emission of $CO_2$. The paper outlines the problems of conventional steam injection and addresses the advantages of superheated steam injection by HTR from the aspects of technology, economy, and environment. A Geographic Information System (GIS) embedded with a thermal hydraulic analysis function is designed and developed to analyze the strategy, which can make the analysis work more practical and credible. Thermal hydraulic analysis using this GIS is carried out by applying this strategy to a reference heavy oil field. Two kinds of injection are considered and compared: wet steam injection by conventional boilers and superheated steam injection by HTR. The heat loss, pressure drop, and possible phase transformation are calculated and analyzed when the steam flows through the pipeline and well tube and is finally injected into the oil reservoir. The result shows that the superheated steam injection from HTR is applicable and promising for thermal recovery of heavy oil reservoirs.

# INTRODUCTION

Steam injection [1–3] is the most widely used and the most efficient process available today among the various enhanced oil recovery techniques currently being employed in heavy oil reservoirs and tar sand deposits. At present, fossil fuel such as oil, coal, and natural gas is burned to provide the energy to steam. Mostly, wet steam is adopted as the injection medium. The relatively large water demand and carbon dioxide ($CO_2$) emissions have emerged as challenges of the steam injection for thermal recovery [4]. Superheated steam, compared with wet steam, has higher heat content and specific volume, by which for steam stimulation not only can the bottom hole steam quality and steam swept volume be increased, but also it is characterized by higher viscosity breaking, heat expansion, blocking, crude oil distillation rate, and oil displacement [5, 6].

The High-Temperature Reactor (HTR) [7–10] has a distinct advantage in inherent safety, economics potential, high efficiency and potential usage for process heat application. The Chinese Module High-Temperature Gas-cooled Reactor (MHTGR), named High-Temperature gas-cooled Reactor-Pebble bed Module (HTR-PM) [11–13], based on the technology and experience of the 10 MW High-Temperature gas-cooled Reactor (HTR-10) [14, 15], will be started in the end of 2012 or later. The demonstration HTR-PM plant has two reactor modules, two steam generators, and one steam turbine which employ superheated steam as the driver power. The helical-coiled once through steam generator can produce the superheated steam. USA utility Entergy completed an economic analysis of using a HTGR (High-Temperature Gas-cooled Reactor) to generate hydrogen and process heat, and also electricity [16]. This study indicates that the cost of process heat or high-temperature steam produced by a HTGR can be competitive with the steam produced by normal fossil plant.

In this paper, the design of the secondary loop after the steam generator of the HTR-PM is modified to fit the need of steam injection for thermal recovery. The superheated steam produced by the steam generator is divided into two branches. The main fluid is applied for steam injection of the heavy oil reservoirs and the other fluid will flow into a two-stage steam jet mixer in order to preheat the inlet water of the steam generator.

In order to analyze the thermal hydraulic behavior of the HTR-PM for heavy oil thermal recovery, a Geographic Information System (GIS) embedded with thermal hydraulic analysis function is designed and developed. Then it is applied to analyze the thermal hydraulic process in a reference heavy oil field where two kinds of injection are considered and compared: wet steam injection by boiler and superheated steam injection by HTR-PM. The heat loss, pressure change, and possible phase transformation are calculated and analyzed when the steam flows through the pipeline and well tube and is finally injected into oil reservoirs. Two kinds of pipeline are considered: overground and underground. Four kinds of oil well structure are included: heat insulation tube with packer, heat insulation tube without packer, normal tube with packer, and normal tube without packer. Calculation results indicate that the superheated steam injection by HTR-PM is applicable and has obviously higher quality of steam stimulation than the wet steam injection by the conventional boiler. Namely, under the same mass flow rate, the steam injection by HTR-PM has characteristics of higher temperature, more enthalpy and larger specific volume than the wet steam injection by conventional boiler. Therefore, it will result in more intensive physical and chemical reactions between reservoir fluid and minerals, higher reduction of oil viscosity, and bigger expansion of crude oil volume, and so forth.

Section 2 of this paper introduces the modified HTR-PM design and strategy when it is applied to heavy oil thermal recovery of an oil field. Section 3 describes the GIS system which is embedded with thermal hydraulic analysis function. Section 4 explains the process and result of the thermal hydraulic analysis of superheated steam injection by HTR and its comparison with wet steam injection by boiler. Section 5describes the conclusions of this paper.

# HTR-PM FOR HEAVY OIL THERMAL RECOVERY

## General Description of HTR-PM

The HTR-PM deploys two pebble-bed modular HTRs each with 250 MW thermal power. Two reactor modules are coupled with two steam generators which are connected to one steam turbine generator with 210 MW electric power. The reactor and the steam generator are installed inside two separate pressure vessels. The pressure vessels are assembled in a staggered, side-by-side arrangement and are connected by a horizontal coaxial hot gas duct. The primary pressure boundary consists of the Reactor Pressure Vessel (RPV), the Steam Generator Pressure Vessel (SGPV), and the Hot gas Duct Pressure Vessel (HDPV), which all are housed in a concrete shielding cavity as shown in Figure 1. The main design parameters are listed in Table 1.

**Table 1**: Major design parameters of HTR-PM [17]

| Item | Value |
|---|---|
| Thermal power (MW) | 2 × 250 |
| Electric power (MW) | 210 |
| Active core diameter (cm) | 300 |
| Equivalent active core height (cm) | 1100 |
| Primary helium pressure (MPa) | 7.0 |
| Average helium temperature at reactor inlet/ outlet (°C) | 250/750 |
| Helium mass flow rate at full power (kgs−1) | 96 |

| Type of steam generator | Once-through helical |
|---|---|
| Main steam flow rate at the inlet of turbine (kgs−1) | 98 |
| Main steam pressure (MPa) | 14.3 |
| Main steam temperature (°C) | 566 |
| Main feed-water temperature (°C) | 205 |

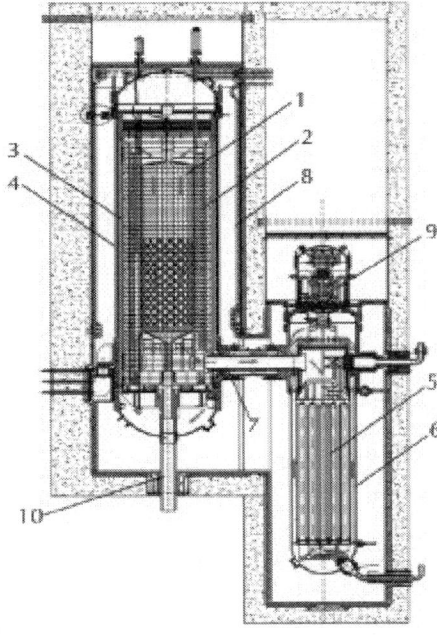

(1) Reflector core

(2) Side reflctor and carbon thermal shield

(3) Core barred

(4) Reflector pressure vessel

(5) Steam generator

(6) Steam generator vessel

(7) Coaxial gas duct

(8) Water-cooling panel

(9) Blower

(10) Feul discharging tube

**Figure 1:** Primary loop of HTR-PM.

As shown in Figure 1, the thermal-hydraulic process related with the reactor and the primary loop of HTR-PM can be mainly explained as follows.

- Before entering the reactor vessel, the helium gas is driven by the helium blower, and the temperature of the inlet gas is around 250°C. After the helium flows through the hot fuel spheres in the reactor core, the temperature of it reaches 750°C. The hot helium exits the reactors and flows through the hot gas duct. The hot helium releases its thermal power to the water in the steam generator where it is cooled down to around 250°C. Then, the helium blower blows the cold helium through the outer coaxial pipes of the hot gas duct. Finally, the cold helium flows into RPV where it is heated again. In this way, a cycle of the helium flow is completed in the primary loop.

- At the side of the secondary loop of HTR-PM plant, the water at a temperature of 205°C in the steam generator is heated to 566°C by hot helium and becomes superheated steam. Then superheated steam with a pressure of 14.3 MPa will drive the steam turbine and its conjoined generator in order to generate electrical power based on the Rankine cycle.

## Status and Existing Problems of Conventional Steam Injection for Thermal Recovery

Steam injection is an established thermal recovery technique that has been applied successfully on many heavy oil reservoirs around the world. The process started in early 1960s with cyclic steam injection. The three main methods of steam injection are Cyclic Steam Stimulation (CSS), Steam Drive (SD), and Steam Assisted Gravity Drainage (SAGD) [2].

The CSS is a simple and cheap method of applying thermal recovery process on a reservoir. It involves injecting steam into a well for several weeks, shutting the well in as long as necessary to allow the steam to heat the oil in the areas around the well, and putting the well back on production to recover the heated oil (Figure 2). This process is repeated when the production from the well declines to a low level. The cycle is repeated many times until the ratio of oil produced to steam injected named Oil-Steam Ratio (OSR) drops to a level that is considered uneconomic. CSS was the first steam injection technique used in heavy oil reservoirs. It can be used as a limited pilot test on a well to investigate whether steam injection can enhance oil recovery in a particular field before undertaking more expensive pattern pilot tests that will require more wells and more equipment.

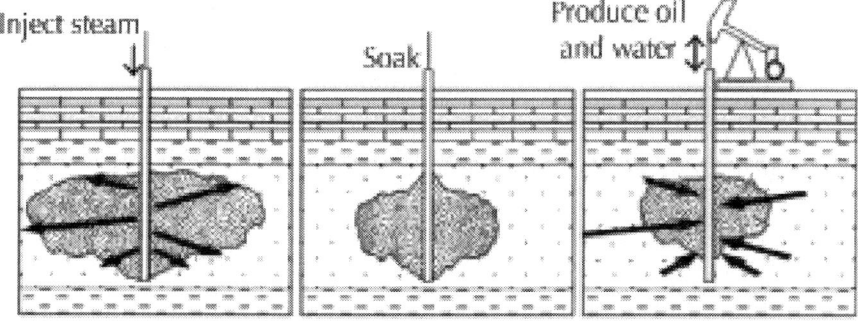

**Figure 2:** Cyclic steam stimulation process.

The scenario of the conventional thermal recovery technology uses the boiler to generate wet steam, which is shown as the bottom part of Figure 3.

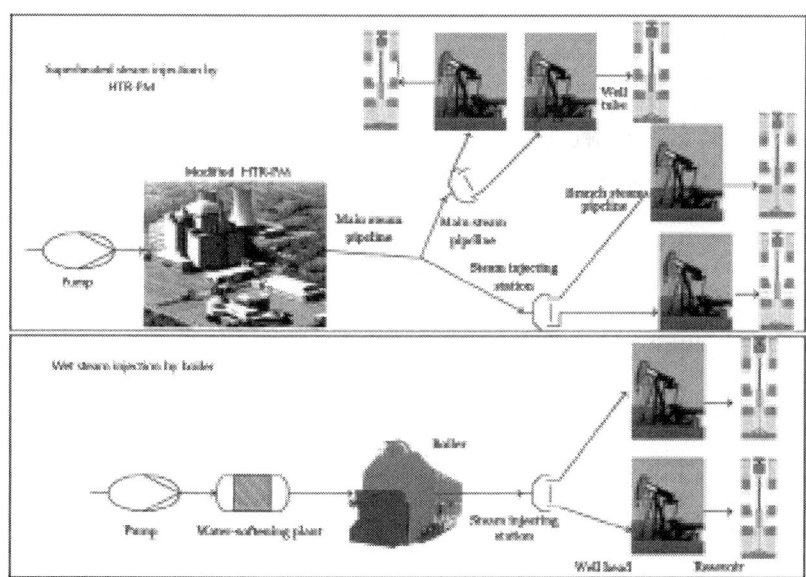

**Figure 3:** Strategies for steam injection by HTR-PM and boiler.

Conventionally, the boiler using fossil fuel such as coal, natural gas, and oil is utilized to heat water to wet steam for thermal recovery of heavy oil reservoirs. The water is driven by a pump and softened by the water-softening plant before entering the boiler. The steam out of the boiler will flow through the pipeline to a steam injecting station near the heavy oil well. Then, the wet steam will pass through the well head to the well tube. Finally, the wet steam will flow into the oil reservoir via the well tube which is usually several hundred meters long.

Because of the limited thermal power of boilers and heat loss during steam transportation, the boilers are built near the heavy oil well to provide enough steam for thermal recovery. One boiler can only provide steam to a limited number of oil wells for steam injection. Usually, there are many boilers in a heavy oil field to provide steam to different heavy oil wells.

The specific enthalpy of the steam or water can be indicated as

$$h(P, T, x) = h'(P) + xQ(P) + x \int_{T_P}^{T} C_P'' \, dt,$$

(1)

Where P is the pressure of steam or water; T is the temperature of steam or water; $x$ is the dryness of steam or water; h (P) is the specific enthalpy of the saturated water at pressure p; Q(P) is the latent heat of vaporization of water at pressure p $C_p$ is the temperature of the saturated steam at pressure p; $T_p$ is the specific heat of the superheated steam at pressure P

The dryness $x$ of wet steam can be expressed as

$$x = \frac{m_s}{m_w},$$

(2)

Where $M_s$ is the mass of the steam component in the wet steam; $m_w$ is the mass of the whole wet steam.

The conventional thermal recovery technology with dispersed steam supply and wet steam injection has the following shortcomings.

- The operation and management of a large number of small boilers is very expensive because of high labor expenditure and relative high price of equipment [18].

- The conventional boiler is burning fossil fuel such as coal, natural gas, and oil to generate the steam, which has an environmental challenge of carbon dioxide emissions and large demand of water [4].

- Loss of heat during steam transportation will reduce the heat capacity of injected steam and obviously worsen its quality since the velocity of flow is relatively slow and the specific enthalpy is low. Therefore, the heavy oil reservoirs cannot gain enough thermal energy even if a lot of water is consumed [5, 6].

- The wet steam has low temperature and low specific volume which result in low stimulation efficiency [5, 6].

## Strategy of Superheated Steam Injection from HTR-PM

The strategy of superheated steam injection by HTR-PM is shown as the top part of Figure 3. Unlike the conventional steam injection method, all superheated steam will be intensively produced by one HTR plant which is located at several kilometers away from the heavy oil field in consideration of safety and engineering requirements. Then, the superheated steam will be delivered all over to the whole heavy oil field through the main steam pipeline with bigger pipe diameter and higher thermal insulation ability. The main steam pipeline will connect to normal pipelines at the diverging points. Finally, the steam will flow into the oil reservoirs through the steam injecting station, well head, and well tube.

For the wet steam, is less than 1 and the third part of (1) is zero? For the superheated steam, is 1? Therefore, the enthalpy difference between the superheated steam and the wet steam at the same pressure can be shown as follows:

$$
h(P, T) - h(P, T_P, x) = (1 - x)Q(P) + \int_{T_P}^{T} C_P'' dt.
$$

(3)

From this equation, it can be found that the superheated steam has much more enthalpy than the wet steam when $\chi$ is not very close to 1 and $T_p - T$ is a big value. The thermal recovery technology by HTR with concentrated steam supply and superheated steam injection is predicted to have the following advantages.

- The operation and management of HTR for steam generation is relatively cheap.

- The steam production by HTR emits less $CO_2$ than the conventional boiler burning fossil fuel.

- Loss of heat during steam transportation will not seriously worsen the quality of injected steam since the velocity of flow is relatively fast and the specific enthalpy is high. Therefore, the heavy oil reservoirs can receive much more thermal energy by superheated steam injection when the same mass of water is injected into oil reservoirs.

- The superheated steam has higher enthalpy and higher specific volume which results in higher quality of steam stimulation [5, 6].

# Adoption of HTR-PM for Heavy Oil Thermal Recovery

The design of the HTR-PM applied to thermal recovery of heavy oil reservoirs is shown in Figure 4. The primary loop and the steam generator are kept as the original design of the HTR-PM in order to take advantage of its mature design. Originally, the secondary loop of the HTR-PM adopts a steam turbine system which generates electricity with a Rankine cycle. A two-stage steam jet mixer system is design to preheat the feed water of the HTR-PM.

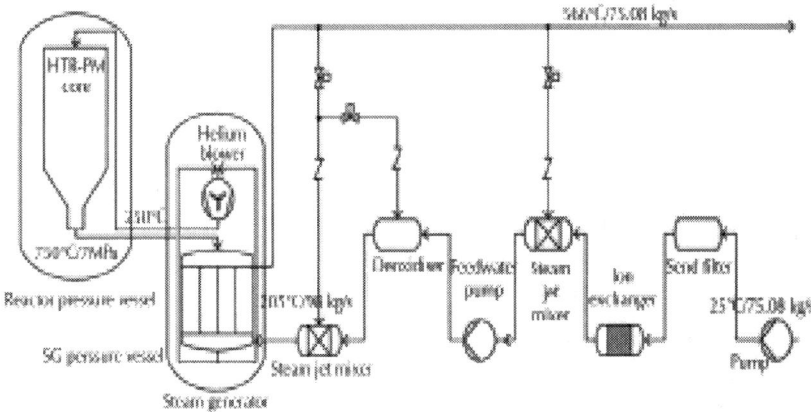

**Figure 4:** Modified design of HTR-PM for thermal recovery.

In the secondary loop of HTR-PM, some basic parameters such as mass flow rate, inlet temperature, and outlet temperature of water should be determined by taking into account the original design of HTR-PM and need of heavy oil thermal recovery. Table 2 shows the basic configuration when a HTR-PM or boilers are applied to thermal recovery of a reference heavy oil field. In order to maintain the safety of HTR-PM and oil field, the HTR-PM is located around 4 km away from the oil field to void the interaction when one of them is in a dangerous status. In this reference oil field, the number of injected wells at the same time is around 30 (20–40) and the mass flow rate of water to each well is around 2.31 kg/s (200 ton/day). A single reactor of HTR-PM can provide enough injected water since the flow out of the steam generator is 98 kg/s, around 8600 ton/day.

**Table 2:** Basic configuration of injection

|  | Injection with boiler/HTR-PM |
|---|---|
| Number of used boiler/HTR-PM reactor at same time | 15/1 |
| Number of injected wells at same time | 30 (20–40) |
| Number of boilers | 100 |
| Number of wells | 6000 |

| | |
|---|---|
| Initial temperature of steam (°C) | 318.05/566 |
| Mass flow out of each boiler/HTR-PM (ton/day) | 200–800/6500 |
| Mass flow rate of water to each well (ton/day) | Around 200 |
| Initial pressure of steam (MPa) | 11/14.3 |
| Initial dryness of steam | 0.725/1 |
| Average length of main stream pipeline (km) | 0/7.5 |
| Area of reference oil field (km2) | 20×20 |
| Average length of branch steam pipeline (km) | 0.8 |
| Average distance from injection station to well head | 0.2 |
| Depth of typical well (m) | 759.75 |

The process in the modified secondary loop of HTR-PM can be described as follows. The water at 75.08 kg/s and 25°C is driven to the secondary loop by a pump. The water is purified and softened by a water treatment plant mainly consisting of several sand filters and several ion exchangers. The cool water is heated by the superheated steam from the output of the steam generator at a steam jet mixer. Then the water is pumped to a deoxidizer and heated again by the superheated steam at another steam jet mixer. Finally, the water at around 205°C and 98 kg/s, which is same as the original design of HTR-PM, is fed to the once-through helical steam generator. The output water of the steam generator at around 566°C is mainly divided in two branches. One fluid is deployed as feedback to the two steam jet mixers and the deoxidizer in order to heat the feed water instead of the conventional electrical heaters. The main fluid at 75.08 kg/s (6500 ton/day) and 566°C is provided to the superheated steam injection system for heavy oil thermal recovery.

## Economic Comparison between Steam Injection by Conventional Boiler and HTR-PM

By considering the construction investment, fuel cost, and operating cost, the full cost of the steam injection by burning coal or natural gas can be estimated [18]. The construction investment and operating cost are converted to cost per ton steam injection which are shown in Table 3. The full cost of steam injection by HTR-PM can also be estimated [19]. In the same way, the construction investment and operating cost of steam injection by HTR-PM are converted to cost per ton of steam injection in the same Table. The full cost of steam injection with unit of $/t means the total cost for injecting 1 ton of steam.

**Table 3:** Costs of steam injection by boiler and HTR-PM

| Injection mode | Construction investment ($/(t·y)) | Fuel cost ($/t or $/m3) | Operating cost ($/t) | Full operating cost ($/t) | Full cost for steam injection ($/t) |
|---|---|---|---|---|---|
| Coal | 29.78 | 27.99 | 8.766 | 12.04 | 14.69 |
| Natural gas | 15.54 | 0.1541 | 10.45 | 14.50 | 15.88 |
| HTR-PM | 95.17 | — | — | 5.62 | 14.07 |

Time = 30 years, bank rate = 8%.

For the uncertainty evaluation, if the bank rate rises to 10% and fuel cost decreases by 10%, the costs can be estimated as shown in Table 4.

**Table 4:** Costs of steam injection at lower fossil fuel price and higher bank rate

| Injection mode | Construction investment ($/(t·y)) | Fuel cost ($/t or $/m3) | Operating cost ($/t) | Full operating cost ($/t) | Full cost for steam injection ($/t) |
|---|---|---|---|---|---|
| Coal | 29.78 | 25.19 | 8.766 | 11.71 | 14.87 |
| Natural gas | 15.54 | 0.1368 | 10.45 | 14.10 | 15.74 |
| HTR-PM | 95.17 | — | — | 5.62 | 15.72 |

Time = 30 years, bank rate = 10%.

From these results, it can be found that the steam injection by HTR-PM is still a competitive choice compared to the steam injection by the conventional boiler burning coal or natural gas even with lower fossil fuel price and higher bank rate.

To consider the quality difference of the steam produced by the fossil fuel boiler or HTR-PM, the full cost can be changed to the enthalpy of injected steam instead of the mass of it. The full costs for steam injection by HTR-PM or burning coal or natural gas are shown in Table 5, which shows the cost to provide 1 kJ enthalpy to the injected steam. According to this table, the steam injection by HTR-PM has an obvious economic advantage on the steam injection by conventional boilers burning coal or natural gas even with lower fossil fuel price and higher bank rate.

**Table 5:** Costs of steam injection by considering enthalpy difference of steam

| Injection mode | Full cost for steam injection (mills/kJ) | Full cost for steam injection under higher bank rate and lower fossil fuel price (mills/kJ) |
|---|---|---|
| Coal | 0.00621 | 0.00629 |
| Natural gas | 0.00672 | 0.00666 |
| HTR-PM | 0.00402 | 0.00449 |

Time = 30 years, higher bank rate = 10%, lower fossil fuel price = 90% of normal price.

# DESIGN AND DEVELOPMENT OF GIS WITH THERMAL HYDRAULIC ANALYSIS FUNCTION

## Need of GIS with Thermal Hydraulic Analysis Function

It is easy to conclude that the concentrative steam supply by HTR has advantages over the conventional dispersed steam supply by boilers at the aspects of labor expenditure [18, 19]. In addition, it can also be predicted that the superheated steam can beat the wet steam because of higher specific enthalpy and specific volume [5, 6]. However, some results from actual engineering side are needed for answering the following questions.

- Will the quality of the steam by HTR be same as or even worse than the quality of the steam by boiler because of heat loss since the superheated steam has to flow through an extra main steam pipeline which is several kilometers long? If it is, the HTR design for thermal recovery technology will lose its most important strength over conventional technology.

- The thermal hydraulic analysis of a whole actual oil field is rather difficult because the situation is very complex when the injected steam flows through the main pipeline, branch pipeline, steam injecting station, well tube, and so forth. The mass flow rate, temperature, pressure, and dryness of the steam change in a nonlinear manner in the process of the steam transportation, which cannot be simply considered as a process in one typical pipeline.

- Since the superheated steam has higher specific volume than the wet steam, the velocity of flow of the superheated steam will be much higher than that of wet steam if the mass rate of injected steam stays the same. Is the pressure drop acceptable or not before the steam flows to oil reservoirs through the extended pipeline?

- Should the main steam pipeline be placed over-ground or underground? Can the superheated steam injection fit for all kinds of well structures?

A GIS with thermal hydraulic analysis function can play the role to answer the above questions. The GIS can serve the following functions.

- Management of geographic information of an oil field such as coordinates of HTR-PM, boiler, pipeline, steam injecting station, and reservoir.

- Management of configuration information of HTR-PM, boiler, pipeline, and reservoir including pressure, temperature, mass flow rate of output steam of HTR-PM or boiler, geometrical and thermodynamics characteristics of pipeline and reservoir, and so forth.

- Calculation of loss of heat, change of pressure, and possible phase change when the steam flows from HTR-PM or boiler to reservoir.

# Structure and Function of the GIS

The structure and function of the GIS are shown in Figure 5. The GIS with thermal hydraulic analysis function is developed based on component-based development where the third-party developed GIS component is embedded in it. It mainly consists of three modules.

- The thermal-hydraulic calculation module can simulate the change of thermal-hydraulic status of the steam from HTR or boiler to oil reservoirs. It can access the information of each geographic symbol in the GIS through temporary input/output file. In the GIS, the geographic symbols represent HTR-PM and various components of the oil field such as boiler, pipeline, steam injecting station, reservoir, and so forth. The thermal-hydraulic calculation module will be explained in Section 4.

- The data treatment module includes three sub-modules: database management module, cache management module, and input/output management module. The database management module can access the basic database, the configuration database, and the result database. The basic database stores the basic information of various geographic symbols in the GIS including name, ID, type, geometry, and so forth. The configuration database stores the information for the thermal hydraulic analysis such as inner diameter of pipeline, outside diameter pipeline, type of pipeline, type of well, heat conductivity of heat-insulating material, and so forth. The result database can save the thermal hydraulic analysis result of each symbol in the GIS. The input/output management module mainly reads and writes the temporary input/output files to provide data to the thermal-hydraulic calculation module and get results from the thermal hydraulic analysis. The cache treatment module gets the information from the database and provides it to the geographic information management module for finding the flow route of the steam out of the HTR or boiler.

- The Geographic Information Management module can realize some basic management functions and advanced management function through the GIS component. The basic management functions can perform the zoom in, zoom out, move, centralize, and other functions for management of the geographic symbol. The advanced management function can determine the flow route of the steam out of the HTR or boiler according to the configuration of each symbol.

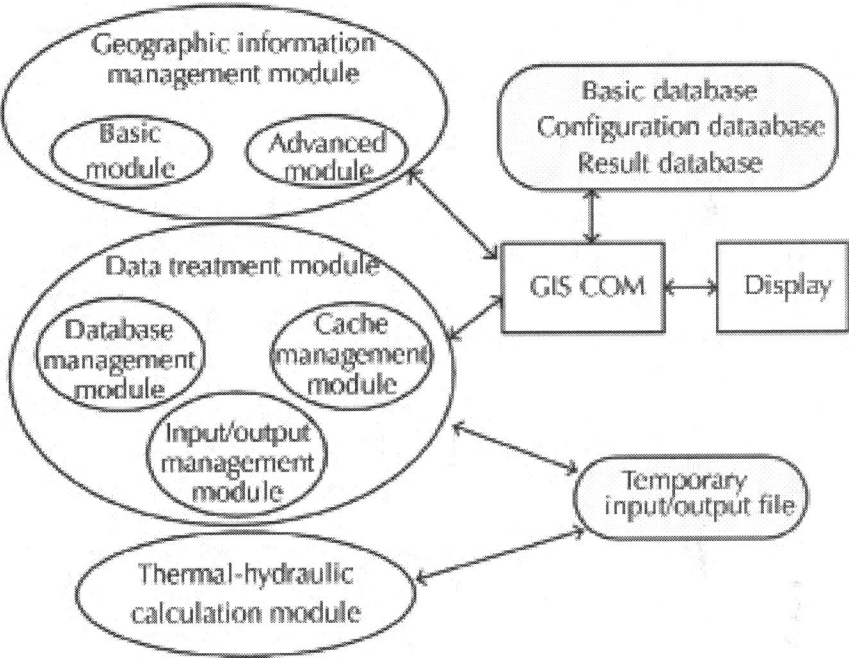

**Figure 5:** Structure and function of the GIS.

## How the GIS Works

The flow chart of the GIS is shown in Figure 6. Firstly, the GIS will load the geographic information and configuration information from the basic and configuration databases. Then, the user can modify the configuration information of the geographic symbol by selecting the corresponding symbol through the GIS interface.

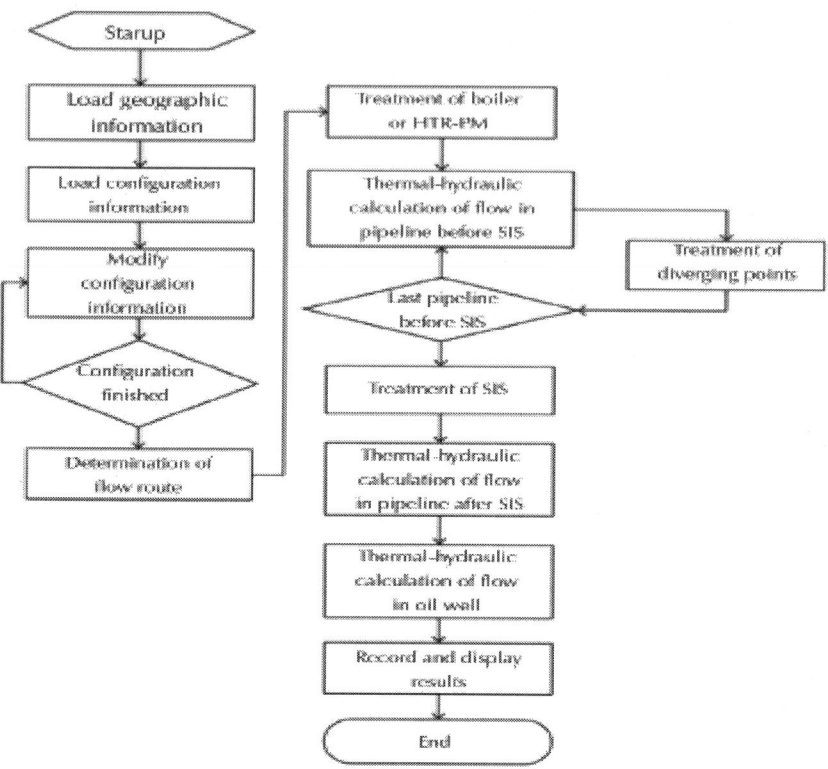

**Figure 6:** Flow chart of the GIS.

After the process of configuration, the flow route from the HTR or boiler to oil reservoirs will be determined according to the configuration since sometime dozens of oil wells out of thousand existing wells in the oil field will be selected for steam injection. The output of the boiler or HTR, such as temperature, pressure, and mass flow rate of steam, will be obtained. Then, the thermal hydraulic analysis will be carried out when the steam flows in the pipeline and diverging points before the Steam Injecting Station (SIS). After the treatment of flow at SIS, the thermal-hydraulic behavior of the flow in the pipeline and the oil well will be calculated. Finally, all the above results will be recorded for display and analysis.

Figure 7 shows the snapshot when the GIS works for thermal hydraulic analysis.

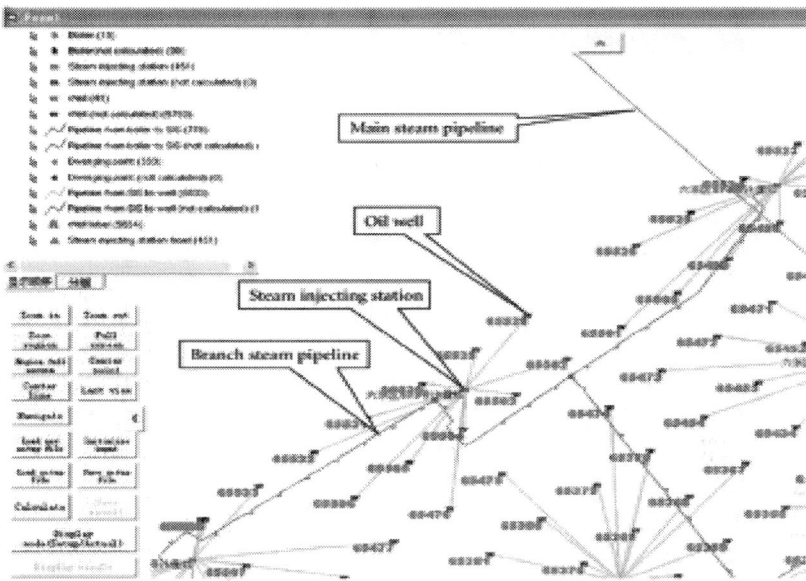

**Figure 7:** Snapshot of the GIS.

# THERMAL-HYDRAULIC MODELS FOR GIS

## Basic Model and Its Treatment

The whole flow route after the HTR or boiler consists of main stream pipeline, diverging point, branch steam pipeline, SIS, and oil well tube. The change of pressure, temperature, and flow rate occurs when the steam flows through the flow route. As boundary conditions, the pressure, temperature (dryness for wet steam), and mass flow rate of the steam are fixed at the outlet of the HTR steam generator or boiler.

When analyzing the thermal-hydraulic behavior of the steam flow in the pipeline and well tube, some basic assumptions are adopted.

- The pressure and flow rate of the steam are analyzed based on quasi-steady state model.

- Steam is compressible and treated as a continuum medium.

According to the assumptions mentioned above, some governing equations can be formulated as follows.

- Mass conservation

$$\frac{u}{v} = \text{const.}$$

(4)

- Momentum conservation

$$g \sin \theta dL + u du + \frac{dP}{\rho} = -\lambda \frac{u^2}{2} \frac{dL}{D_i}.$$

(5)

- Energy conservation

$$dQ = g \sin \theta dL + dh + u\, du.$$

6

(6)

In (4)–(6), $u$ is the flow speed of steam or water; $v$ is the specific volume of steam or water;   is the density of steam or water; $P$ is the pressure of steam or water; $L$ is the length of the steam or water flow;   is the friction resistance coefficient of the pipeline or well tube; $Di$ is the equivalent diameter of the pipeline or well tube; $g$ is the acceleration of gravity;   is the angle between the flow and the horizontal direction; $dQ$ is the heat absorbed per unit mass of steam or water; $h$ is the specific enthalpy of the steam or water.

 can be calculated as follows [20]:

$$\lambda = \frac{64}{\text{Re}} \qquad (\text{Re} < 2300)$$

$$\lambda = \frac{0.316}{\text{Re}^{0.25}} \qquad \left( 2300 \le \text{Re} < 80 \frac{D_i}{\Delta_d}, \ \text{Re} < 8 \times 10^4 \right)$$

$$\frac{1}{\sqrt{\lambda}} = 2\lg\left(\mathrm{Re}\,\sqrt{\lambda}\right) - 0.8$$

$$\left(2300 \le \mathrm{Re} < 80\frac{D_i}{\Delta_d}, \quad \mathrm{Re} \ge 8 \times 10^4\right)$$

$$\frac{1}{\sqrt{\lambda}} = -2\lg\left(\frac{\Delta_d}{3.7D_i} + \frac{2.51}{\mathrm{Re}\,\sqrt{\lambda}}\right)$$

$$\left(80\frac{D_i}{\Delta_d} \le \mathrm{Re} < 4160\left(\frac{D_i}{2\Delta_d}\right)^{0.85}\right)$$

$$\lambda = \left(2\lg\left(\frac{D_i}{2\Delta_d}\right) + 1.74\right)^{-2} \quad \left(\mathrm{Re} \ge 4160\left(\frac{D_i}{2\Delta_d}\right)^{0.85}\right), \tag{7}$$

where Re is the Reynolds number; $_d$ is roughness of tube wall. In the three conservation equations (4)–(6) in this model, the independent variables are pressure $P$ and temperature $T$ (dryness $x$ for wet steam). The related differential equations are described by

$$\frac{dP}{dL} = f_1(P, T) \quad \text{or} \quad \frac{dP}{dL} = f_1'(P, x)$$

$$\frac{dT}{dL} = f_2(P, T) \quad \text{or} \quad \frac{dx}{dL} = f_2'(P, x). \tag{8}$$

In order to get the above two equations, from the three conservation equations, the following expression of enthalpy and specific volume are employed:

$$\frac{dh}{dL} = \frac{\partial h}{\partial P}\frac{dP}{dL} + \frac{\partial h}{\partial T}\frac{dT}{dL} \quad \text{or} \quad \frac{dh}{dL} = \frac{\partial h}{\partial P}\frac{dP}{dL} + \frac{\partial h}{\partial x}\frac{dx}{dL} \tag{9}$$

$$\frac{dv}{dL} = \frac{\partial v}{\partial P}\frac{dP}{dL} + \frac{\partial v}{\partial T}\frac{dT}{dL} \quad \text{or} \quad \frac{dv}{dL} = \frac{\partial v}{\partial P}\frac{dP}{dL} + \frac{\partial v}{\partial x}\frac{dx}{dL}. \tag{10}$$

From (4), (5), (6), (9), and (10), (11) and (12) are deduced for superheated steam or water (use $x$ instead of $T$ for wet steam) as follows:

$$\frac{dP}{dL} = \frac{BF + CE}{AE - BD} \tag{11}$$

$$\frac{dT}{dL} = \frac{AF + CE}{BD - AE} \quad \text{or} \quad \frac{dx}{dL} = \frac{AF + CE}{BD - AE}, \tag{12}$$

Where

$$A = v + \frac{u^2}{v}\frac{\partial v}{\partial P},$$

$$B = \frac{u^2}{v}\frac{\partial v}{\partial T} \quad \text{or} \quad B = \frac{u^2}{v}\frac{\partial v}{\partial x},$$

$$C = g\sin\theta + \lambda\frac{u^2}{2D_i},$$

$$D = \frac{\partial h}{\partial P} + \frac{u^2}{v}\frac{\partial v}{\partial P},$$

$$E = \frac{\partial h}{\partial T} + \frac{u^2}{v}\frac{\partial v}{\partial T} \quad \text{or} \quad E = \frac{\partial h}{\partial x} + \frac{u^2}{v}\frac{\partial v}{\partial x},$$

$$F = \frac{dQ}{dL} - g\sin\theta. \tag{13}$$

Equations (11) and (12) are numerically solved by using a variable step four-order explicit Runge-Kutta method [21, 22]. For the superheated steam or water, the iteration equation (for the wet steam, use $x$ instead of $T$) can be expressed as follows:

$$P' = P_0 + \frac{l}{6}(K_{11} + 2K_{21} + 2K_{31} + K_{41}),$$

$$T' = T_0 + \frac{l}{6}(K_{12} + 2K_{22} + 2K_{32} + K_{42}),$$

$$K_{11} = \left.\frac{dP}{dL}\right|_{P=P_0,T=T_0},$$

$$K_{12} = \left.\frac{dT}{dL}\right|_{P=P_0,T=T_0},$$

$$K_{21} = \left.\frac{dP}{dL}\right|_{P=P_0+(l/2)K_{11},T=T_0+(l/2)K_{12}},$$

$$K_{22} = \left.\frac{dT}{dL}\right|_{P=P_0+(l/2)K_{11},T=T_0+(l/2)K_{12}},$$

$$K_{31} = \left.\frac{dP}{dL}\right|_{P=P_0+(l/2)K_{21},T=T_0+(l/2)K_{22}},$$

$$K_{32} = \left.\frac{dT}{dL}\right|_{P=P_0+(l/2)K_{21},T=T_0+(l/2)K_{22}},$$

$$K_{41} = \left.\frac{dP}{dL}\right|_{P=P_0+(l/2)K_{31},T=T_0+(l/2)K_{32}},$$

$$K_{42} = \left.\frac{dT}{dL}\right|_{P=P_0+(l/2)K_{31},T=T_0+(l/2)K_{32}}, \tag{14}$$

where $l$ is the iteration length of $L$; $P_0$ and $T_0$ are the value of the previous iteration step.

## Calculation of Heat Loss

The structure of overground or underground pipeline is shown in Figure 8. The structure of the heat insulation oil well tube is shown in Figure 9. There is a packer at the bottom of the whole well tube near the oil layer which is used to seal the cavity inside the sleeve. With the packer, the cavity will be filled with air only. Otherwise, water will occupy a partial height of the cavity. Different from the heat insulation well tube, the normal well tube does not have the heat barrier material and the thermal sleeve of Figure 9.

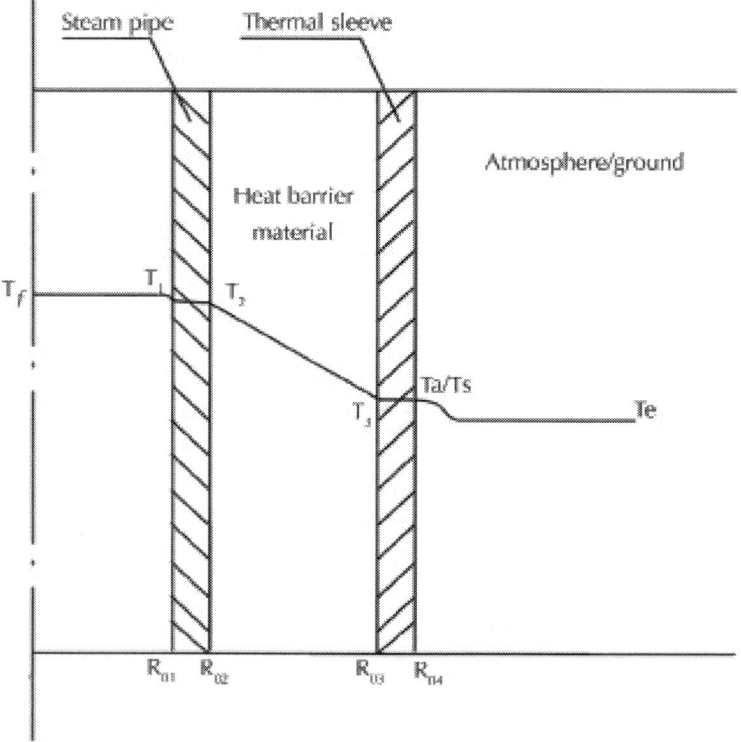

**Figure 8**: Structure of overground or underground pipeline.

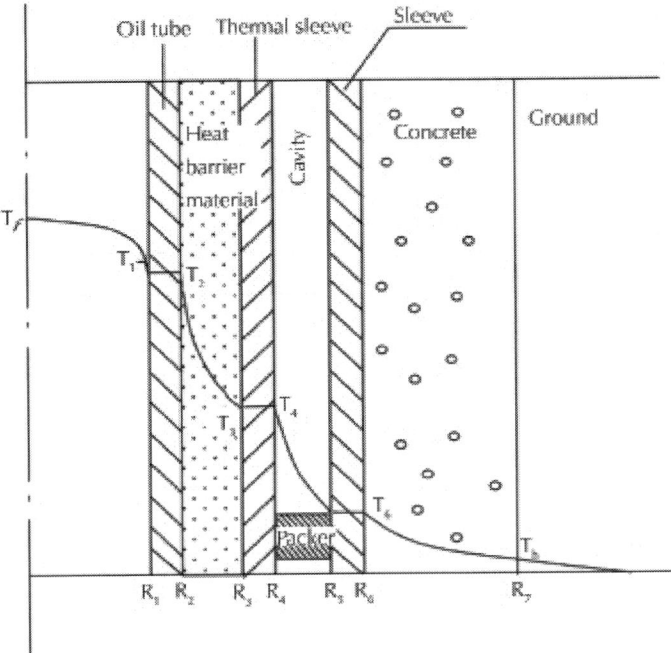

**Figure 9**: Structure of insulation oil well tube.

When solving the equations mentioned in Section 4.1, the heat absorbed by the pipeline wall and the well tube wall should be obtained [23].

For the case of heat which is lost from the steam or water through the overground pipeline wall to the atmosphere as follows:

$$dQ_o = 2\pi R_{o4} \times U_o \left( T_f - T_a \right) dL,$$

(15)

where $R_{o4}$ is the outer radius of the thermal sleeve; $U_o$ is the equivalent coefficient of heat transfer; $T_f$ is the temperature of steam or water fluid; $T_a$ is the temperature of atmosphere next to the thermal sleeve.

$U_o$ can be described as follows, in which the thermal resistance of convective heat transfer between steam/water and pipe and the thermal resistance of steam pipe and thermal sleeve are ignored:

$$U_o = \left[ \frac{R_{o4} \ln(R_{o3}/R_{o2})}{\lambda_{ins0}} + \frac{1}{h_{c0}} \right]^{-1}, \tag{16}$$

where $R_{o2}$ is the outer radius of the steam pipe; $R_{o3}$ is the inner radius of the thermal sleeve; $\lambda_{ins0}$ is the coefficient of thermal conductivity of the heat barrier material; $h_{c0}$ is the coefficient of convective heat transfer of air out of the thermal sleeve.

For the case of heat which is lost from the steam or water through the underground pipeline wall to the earth

$$dQ_u = 2\pi R_{o4} \times U_u \left( T_f - T_s \right) dL, \tag{17}$$

where $U_u$ is the equivalent coefficient of heat transfer; $T_s$ is the temperature of soil next to the thermal sleeve.

$U_u$ can be described as follows:

$$U_u = \left[ \frac{R_{o4} \ln(R_{o3}/R_{o2})}{\lambda_{ins0}} \right]^{-1}. \tag{18}$$

For the case of heat which is lost from steam or water through the oil well tube wall to the earth

$$dQ_t = 2\pi R_7 \times U_t \left( T_f - T_h \right) dL, \tag{19}$$

where $R_7$ is the outer diameter of the concrete; $U_t$ is the equivalent coefficient of heat transfer; $T_h$ is the temperature of soil next to the concrete.

$U_t$ can be described as the following kinds of situation in which the thermal resistance of convective heat transfer between steam/water and pipe and the thermal resistance of the steam pipe, thermal sleeve and sleeve are ignored.

- Heat insulation tube with packer

$$U_t = \left[ \frac{R_7}{R_2 U_{ins}} + \frac{R_7}{R_4(h_c + h_r)} + \frac{R_7 \ln(R_7/R_6)}{\lambda_{cem}} \right]^{-1}.$$

(20)

- Heat insulation tube without packer
  When the cavity is filled with air, it is the same as (20).
  When the cavity is filled with water as follows:

$$U_t = \left[ \frac{R_7}{R_2 U_{ins}} + \frac{R_7}{R_4 h_c} + \frac{R_7 \ln(R_7/R_6)}{\lambda_{cem}} \right]^{-1}.$$

(21

- Normal tube with packer

$$U_t = \left[ \frac{R_7}{R_4(h_c + h_r)} + \frac{R_7 \ln(R_7/R_6)}{\lambda_{cem}} \right]^{-1}.$$

(22)

- Normal tube without packer
  When the cavity is filled with air, it is the same as (22).
  When the cavity is filled with water as follows:

$$U_t = \left[ \frac{R_7}{R_4 h_c} + \frac{R_7 \ln(R_7/R_6)}{\lambda_{cem}} \right]^{-1},$$

(23)

where $R_2$, $R_4$, $R_6$ and, $R_7$ are the outer radius of oil tube, outer radius of the thermal sleeve, outer radius of the sleeve, and outer radius of concrete, respectively; $U_{ins}$ is the equivalent coefficient of heat transfer of the heat insulation tube; $h_c$ is the equivalent coefficient of conductive and convective heat transfer in the cavity; $h_r$ is the coefficient of radiation heat transfer in the cavity; $\lambda_{cem}$ is the coefficient of conductive heat transfer of the concrete.

## Calculation of Local Pressure Change

For the case of superheated steam or water, the local pressure change can be described as follows:

$$\Delta p = \lambda_l \frac{u_d^2}{2v},$$

(24)

where $\lambda_l$ is the equivalent local resistance coefficient; $u_d$ is the average flow rate of downstream steam or water.

The value of $\lambda_l$ can be divided as the four following situations which are shown in Table 6 [20]:

**Table 6**: Values of $\lambda_l$ in the case of steam or water

| Situation | | Value |
|---|---|---|
| Sudden enlargement pipe or tube | | $\lambda_l = \left( \dfrac{A_2}{A_1} - 1 \right)^2$ <br><br> Where A1 is the sectional area of upstream flow channel; <br><br> A2 is the sectional area of downstream flow channel. |
| Sudden contraction pipe or tube | | $\lambda_l = 0.5 \left( 1 - \dfrac{A_2}{A_1} \right)$ |
| 90° bend | | $\lambda l = 0.12$ |

| Valve | Gate valve | $\lambda l = 0.2$ |
|---|---|---|
|  | Ball valve | $\lambda l = 10.0$ |
|  | Control valve | $\lambda l = 5.0$ |

For the case of wet steam, Δp can be obtained by treating the wet steam as the single phase mentioned above. Then, the local pressure change will be corrected as follows [20].

$$\Delta p_s = \Delta p \left( 1 + \frac{B}{X} + \frac{1}{X^2} \right), \tag{25}$$

Where

$$X = \left( \frac{1-x}{x} \right)^{0.9} \cdot \left( \frac{v'}{v''} \right)^{0.5} \cdot \left( \frac{\mu'}{\mu''} \right)^{0.1},$$

$$B = B_1 \cdot \left( \sqrt{\frac{v''}{v'}} + \sqrt{\frac{v'}{v''}} \right), \tag{26}$$

where $v'$ is the specific volume of saturated water; $v''$ is the specific volume of saturated steam; $\mu'$ is the dynamic viscosity coefficient of saturated water; $\mu''$ is the dynamic viscosity coefficient of saturated steam.

The value of $B_1$ can be divided as the four following situations which are shown in Table 7.

**Table 7:** Values of $B_1$ in the case of wet steam

| Situation | | Value |
|---|---|---|
| Sudden enlargement pipe or tube | | B1= 1.0 |
| Sudden contraction pipe or tube | | B1= 1.0 |
| 90° bend | | $B_1 = 1 + 35\dfrac{D_i}{l}$ <br><br> where l is the length of the bend |
| Valve | Gate valve | B1= 1.5 |
| | Ball valve | B1= 2.3 |
| | Control valve | B1= 1.0 |

# THERM1AL HYDRAULIC ANALYSIS FOR THERMAL RECOVERY

## Basic Configuration

Table 8 shows the basic parameters and their values of the main steam pipeline and branch steam pipeline. There are four kinds of branch steam pipeline with different inner diameter and thickness according to their mass flow rate. According to the study, the thermal conductivity of heat barrier material is changed with the temperature of fluid [24]. When the temperature of the fluid is higher than 400°C, a linear relation is adopted in which the thermal conductivity will be increased with the rise of temperature of the fluid. Otherwise, the thermal conductivity is set as a constant. The thermal conductivity of the main steam pipeline is less than that of the branch steam pipeline.

**Table 8**: Basic parameters of pipeline

| Parameter | Value |
|---|---|
| Roughness (mm) | 0.014 |
| Inner diameter (m) | 0.4/0.3/0.2/0.1 |
| Thickness (mm) | 15/15/8/7 |
| Thermal conductivity coefficient of heat | $0.058 + t \times 0.9E - 4$ (t > 400°C) |
| barrier material (J/($m \cdot s \cdot K$)) | 0.094/0.125 (t ≤ 400°C) |
| Thickness of heat barrier material (m) | 0.13 |
| Horizontal angle (°) | 0 |
| Position | Overground/underground |
| Height of bracket (m) | 0.6/0.3 |
| Cross-sectional area of bracket in thermal sleeve (m2) | 0.004 |

Table 9 shows the basic parameters and their values of oil well. Here, the values of basic parameters of all oil wells are set as the same value in order to compare the difference of heat loss, pressure change, and possible phase transformation of different designs. Actually, the values of basic parameters of different oil wells may be different to each other, such as height of well, roughness of well tube, and so forth. The effect of the joint is considered which will increase the heat loss when the well tube is the heat insulation tube.

**Table 9**: Basic parameters of oil well

| Parameter | Value |
|---|---|
| Roughness of well tube (mm) | 0.014 |
| Inner diameter of well tube (mm) | 62 |
| Outer diameter of well tube (mm) | 73 |
| Thermal conductivity coefficient of heat barrier material (J/($m \cdot s \cdot K$)) | $0.0443 + t \times 1.39E - 4$ |
| Thickness of thermal sleeve (mm) | 19.5 |
| Inner blackness of cavity | 0.9 |

| | |
|---|---|
| Outer blackness of cavity | 0.56 |
| Inner diameter of sleeve (mm) | 224.4 |
| Outer diameter of sleeve (mm) | 324.4 |
| Thickness of concrete (mm) | 30 |
| Thermal conductivity coefficient of concrete $(J/(m \cdot s \cdot K))$ | 0.516 |
| Length of each well tube (m) | 10 |
| Length of each joint (m) | 0.13 |
| Inner diameter of joint (mm) | 60 |
| Outer diameter of joint (mm) | 76 |
| Height of well (m) | 759.75 |

Table 10 shows the basic parameters and their values of environment.

**Table 10**: Parameter of environment

| Parameter | Value |
|---|---|
| Tundra exists | NO |
| Temperature of atmosphere and earth surface (°C) | 20 |
| Starting temperature of zone of constant temperature (°C) | 15 |
| Starting depth of zone of constant temperature (m) | 30 |
| Temperature gradient of zone of constant temperature (°C/m) | 0.018 |
| Wind speed of earth surface (m/s) | 2 |
| Thermal conductivity of soil except tundra (J/(msK)) | 2.6528 |
| Thermal diffusivity of soil except tundra (m2/s) | 9.972E – 07 |

# Introduction of Result

The GIS embedded with thermal hydraulic analysis function is applied to analyze the thermal hydraulic process in the reference heavy oil field whose basic configuration is mentioned above. The heat loss, pressure change, and possible phase transformation are calculated and analyzed when the steam flows through the pipeline and is injected into oil reservoirs.

Table 11 shows the different scenarios and its serial number. Two kinds of injection are considered: wet steam injection by boiler and superheated steam injection by HTR-PM. Two kinds of pipeline are considered: overground and underground. Four kinds of oil reservoir structure are included: heat insulation tube with packer, heat insulation tube without packer, normal tube with packer, and normal tube without packer.

**Table 11**: Different scenarios and their serial numbers

|  | HTR/PM overground | HTR-PM/ underground | Boiler/ overground | Boiler/ underground |
|---|---|---|---|---|
| Normal tube without packer | 1 | 5 | 9 | 13 |
| Heat insulation tube without packer | 2 | 6 | 10 | 14 |
| Normal tube with packer | 3 | 7 | 11 | 15 |
| Heat insulation tube with packer | 4 | 8 | 12 | 16 |

Figure 10 shows the heat exchange of the 16 scenarios at different stages when the steam flows out of HTR-PM or boilers to oil reservoirs. The loss of heat is an average value since 30 oil wells are selected for injection and the loss of heat will be different when the steam flows to different oil wells. For the first 8 scenarios where superheated steam is provided by HTR-PM, the enthalpy of the fluid out of HTR-PM is around 3500 kJ/kg. After flowing through the main steam pipeline, the enthalpy of the fluid before entering the steam line goes down to 2800–2900 kJ/kg. Because of the loss of heat in the branch steam pipeline, the enthalpy of the fluid before the well head becomes less than 2800 kJ/kg. The heat loss in the well tube is determined by the structure of the well, where the biggest loss occurs in the normal tube without packer and the smallest loss in the heat insulation tube with packer. For the other 8 scenarios where the wet steam is provided by boiler, the enthalpy of the fluid out of boilers is less than 2400 kJ/kg. Since the fluid from the boiler does not flow through the main steam pipeline but directly flows into the branch steam pipeline, there are only three stages for the fluid provided by boiler. The features of heat loss of the fluid in the last 8 scenarios are the same as the first 8 scenarios.

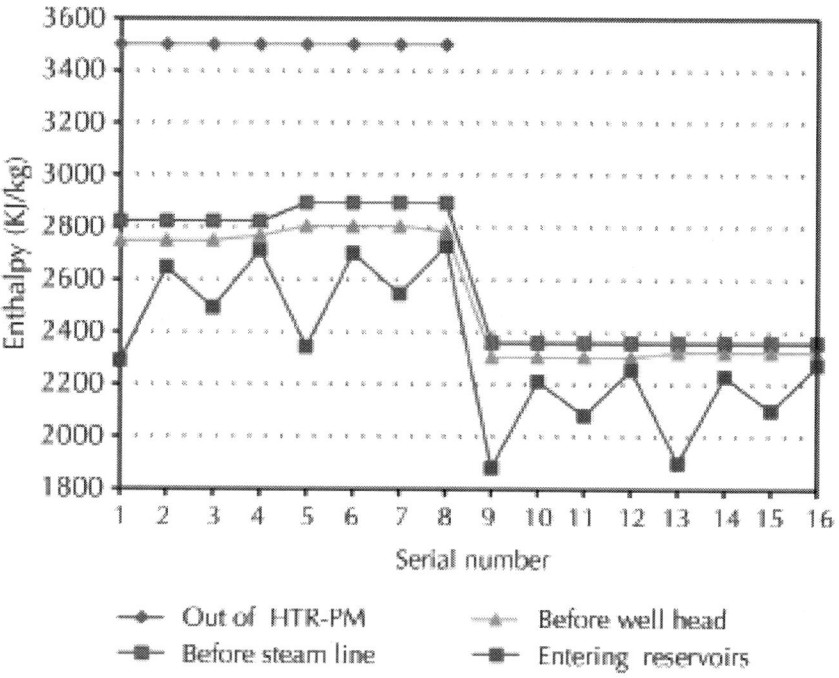

**Figure 10**: Heat of steam at different stages in the 16 scenarios.

Figure 11 shows the dryness of the fluid entering the oil reservoirs and the pressure change of the fluid after the steam flows through the main steam pipeline, the branch steam pipeline, and the well tube. The dryness and the pressure change are also average values since their values will be different when the steam flows to the 30 different oil wells. The dryness of the fluid entering reservoirs of scenarios 1–8 is much higher than that of scenarios 9–16. Especially for scenarios 2, 4, 6, 8 where the heat insulation tube is used, almost all oil reservoirs are injected by superheated steam (29 out of 30 wells). The pressure change of fluid of scenarios 1–8 is much higher than that of scenarios 9–16 since the fluid out of HTR-PM have to flow through an extra main steam pipeline which is several kilometers long.

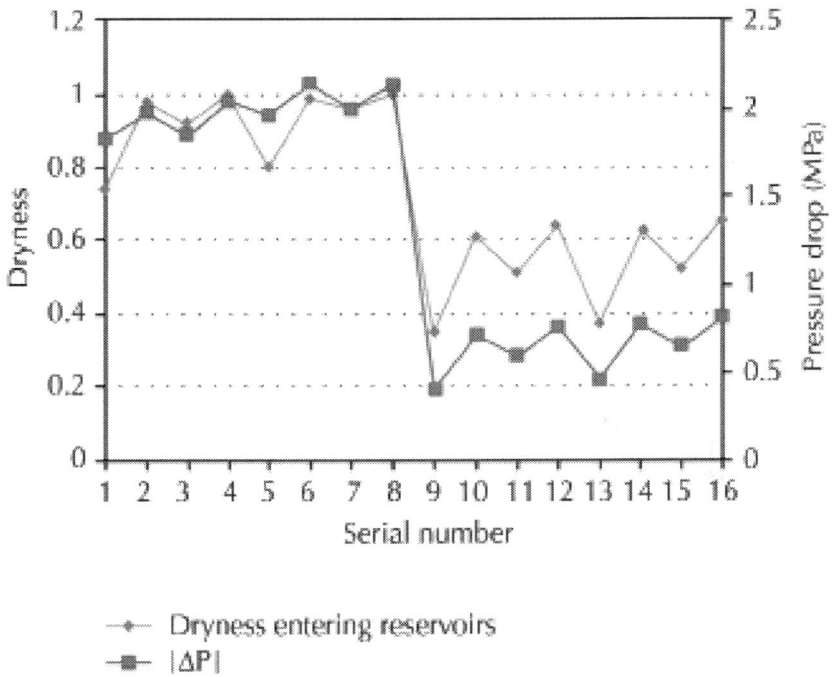

**Figure 11**: Dryness and pressure drop of injected steam in the 16 scenarios.

# DISCUSSION

- For all 8 scenarios where the steam is provided by HTR-PM, the majority of the heat loss exists in the main steam pipeline since it is usually several kilometers long. Another reason is that the temperature of the fluid in the main steam pipeline is relatively higher than the temperature of the fluid in other stages, such in the branch steam pipeline and well tube.

• The heat loss of the fluid in scenarios 1–8 is much higher than that in scenarios 9–16. However, the heat of fluid entering the oil reservoirs in scenarios 1–8 is much higher than that in scenarios 9–16.

• In scenarios 5–8 and 13–16 where the pipeline is underground, the heat loss is less than that in scenarios 1–4 and 9–12 when the pipeline is overground. However, the difference between them is small. It shows that the underground pipeline has no obvious advantage on heat insulation over the overground pipeline under the present condition.

• From well head to oil reservoirs, the heat loss in the heat insulated well tube is much lower than that in normal tube.

• The dryness of the fluid entering the oil reservoirs in scenarios 1–8 is much higher than that in scenarios 9–16. Especially in scenarios 2, 4, 6, and 8 where the insulation well tube is used, the fluid entering the majority of 30 oil reservoirs is superheated steam.

• In scenarios 1–8 where steam is provided by HTR-PM, the total flow resistance is much higher than that in scenarios 9–16 where the steam is provided by boiler. The main reason is that the fluid out of HTR-PM has to flow through an extra main steam pipeline which is several kilometers long. Another reason is that the specific volume of the fluid in scenarios 1–8 is relatively higher than that in scenarios 9–16.

# CONCLUSIONS

In this paper, the strategy of superheated steam injection with a modified design of the secondary loop of HTR-PM is proposed for thermal recovery of heavy oil. A GIS embedded with thermal hydraulic analysis function is developed to analyze the thermal hydraulic parameters of this strategy. The present design of one single reactor of the HTR-PM can fit the need of thermal recovery of the reference oil field where around 20–40 oil wells out of several thousand wells are injected with steam at the same time. Of course, the number of reactors can be changed according to the steam requirements of other oil fields. Thermal hydraulic analysis on two kinds of pipelines and four kinds of well tubes in the reference heavy oil field shows that the strategy is applicable and promising(i)Thermal recovery by HTR-PM is superior to that by conventional boilers since the heat and dryness of the fluid entering the oil reservoirs is much higher than that provided by conventional boilers, which can improve the quality of steam stimulation for heavy oil recovery.(ii)The location of main pipeline (overground or underground) has no obvious influence on quality of injecting steam. However, the overground main steam pipeline is recommended on considerations of installation and maintenance.

# ACKNOWLEDGMENTS

This work has been supported by the Chinese National Science and Technology Major Project (Grant no. ZX06901 and ZX06908).

# REFERENCES

1. K. L. Goyal and S. Kumar, "Chapter 11 steamflooding for enhanced oil recovery," Developments in Petroleum Science, vol. 17, pp. 317–349, 1989. · ·

2. E. Nnaemeka, Petroleum Reservoir Engineering Practice, Prentice Hall, Boston, Mass, USA, 2010.

3. K. F. Johannes, Petroleum Engineer›s Guide to Oil Field Chemicals and Fluids, Elsevier, Oxford, UK, 2012.

4. A. R. Kovscek, "Emerging challenges and potential futures for thermally enhanced oil recovery,"Journal of Petroleum Science and Engineering. In press.

5. W. Xianghong, X. Anzhu, and F. Hailiang, "An integrated evaluation on factors affecting the performance of superheated steam huff and puff in heavy oil reservoirs," Petroleum Exploration and Development, vol. 37, no. 5, pp. 608–613, 2010. · ·

6. T. Zhou, L. Cheng, C. He, Z. Pang, and F. Zhou, "Calculation model of on-way parameters and heating radius in the

superheated steam injection wellbore," Petroleum Exploration and Development, vol. 37, no. 1, pp. 83–88, 2010.

7.    E. Baust and I. A. Weisbrodt, "Introducing the high-temperature reactor into the market-status and strategy," Nuclear Engineering and Design, vol. 121, no. 2, pp. 311–315, 1990.

8.    G. Brinkmann, J. Pirson, S. Ehster et al., "Important viewpoints proposed for a safety approach of HTGR reactors in Europe: final results of the EC-funded HTR-L project," Nuclear Engineering and Design, vol. 236, no. 5-6, pp. 463–474, 2006.

9.    R. Kuhr, "HTR›s role in process heat applications," Nuclear Engineering and Design, vol. 238, no. 11, pp. 3013–3017, 2008.

10.   R. Reimert and M. Schad, "Process heat from modularized HTR," Nuclear Engineering and Design, vol. 251, pp. 244–251, 2012.

11.   Z. Zhang and Y. Sun, "Economic potential of modular reactor nuclear power plants based on the Chinese HTR-PM project," Nuclear Engineering and Design, vol. 237, no. 23, pp. 2265–2274, 2007.

12.   Z. Zhang, Z. Wu, D. Wang et al., "Current status and technical description of Chinese 2 × 250 MWth HTR-PM demonstration plant," Nuclear Engineering and Design, vol. 239, no. 7, pp. 1212–1219, 2009.

13.   Y. Dong and Z. Gao, "Thermal-hydraulic feasibility analysis on uprating the HTR-PM," Nuclear Engineering and Design, vol. 236, no. 5-6, pp. 510–515, 2006.

14.   Z. Wu, D. Lin, and D. Zhong, "The design features of the HTR-10," Nuclear Engineering and Design, vol. 218, no. 1–3, pp. 25–32, 2002.

15.   Y. Xu and K. Zuo, "Overview of the 10 MW high temperature gas cooled reactor—test module project," Nuclear Engineering and Design, vol. 218, no. 1–3, pp. 13–23, 2002.

16.   W. Dalrymple, "Beyond electricity," Nuclear Engineering International, vol. 54, no. 660, pp. 34–35, 2009.

17.   Y. Zheng, L. Shi, and Y. Dong, "Thermohydraulic transient studies of the Chinese 200 MWe HTR-PM for loss of forced cooling accidents," Annals of Nuclear Energy, vol. 36, no. 6, pp. 742–751, 2009.

18. J. Zhang, X. Jia, G. Zeng, Q. Wei, J. Yuan, and R. Liu, "Economic optimization of full life-circle for heavy oil thermal recovery by steam injection process in Karamay oilfield," Xinjiang Pretrolium Geology, vol. 33, no. 1, pp. 80–81, 2012.

19. L. Dong, "Modeling and analysis of modular high-temperature gas-cooled reactor super-critical unit base on vPower simulation platform," Tsinghua University, pp. 167–173, 2012.

20. J. Liu, "Research on improving steam thermal efficiency," South West Petroleum University of China, 2003.

21. J. Ye, Y. Zhou, X. Chen, Y. Ma, F. Li, and Y. Dong, "Simulation of a once-through, helical-coiled steam generator of high temperature gas-cooled reactor," in Proceedings of the 17th International Conference on Nuclear Engineering, (ICONE17 ‹09), pp. 645–650, Brussels, Belgium, July 2009.

22. M. Wu, "Analysis and simulation of the thermal-hydraulic processes of primary loop of HTGR,"Tsinghua University, pp. 24–39, 2008.

23. Y. Zeng, X. Li, L. Chen, X. Lu, and Q. Wang, "Analysis of the wellbore heat loss in development heavy oil reservoir by steam injection," Drilling & Production Technology, vol. 29, no. 4, pp. 44–46, 2006.

24. X. Liu, M. Li, and Y. Jian, "Simulative experimental research on the thermal insulation construction of the gas injecting pipeline using thick oil heat extracting," Pipeline Technique and Equipment, no. 2, pp. 8–10, 2000.

# Chapter 2

# Thermodynamic Constraints on Methanogenic Crude Oil Biodegradation

Jan Dolfing[1], Stephen R Larter[2], and Ian M Head[1]

[1]School of Civil Engineering and Geosciences, Newcastle University, Newcastle upon Tyne, UK

[2]Petroleum Reservoir Group, Department of Geoscience and Alberta Ingenuity Center for In Situ Energy, University of Calgary, Calgary, Alberta, Canada

## ABSTRACT

Methanogenic degradation of crude oil hydrocarbons is an important process in subsurface petroleum reservoirs and anoxic environments contaminated with petroleum. There are several possible routes whereby hydrocarbons may be converted to methane: (i) complete oxidation of alkanes to $H_2$ and $CO_2$, linked to methanogenesis from $CO_2$ reduction; (ii) oxidation of alkanes to acetate and $H_2$, linked to acetoclastic methanogenesis and $CO_2$ reduction; (iii) oxidation of alkanes to acetate and $H_2$, linked to syntrophic acetate oxidation and methanogenesis from $CO_2$ reduction; (iv) oxidation of alkanes to acetate alone, linked to acetoclastic methanogenesis and (v) oxidation of alkanes to acetate alone, linked to syntrophic acetate oxidation and methanogenesis from $CO_2$ reduction. We have developed the concept of a 'window of opportunity' to evaluate the range of conditions under which each route is thermodynamically feasible. On this basis the largest window of opportunity is presented by the oxidation of alkanes to acetate alone, linked to acetoclastic methanogenesis. This contradicts field-based evidence that indicates that in petroleum rich environments acetoclastic methanogenesis is inhibited and that methanogenic $CO_2$ reduction is the predominant methanogenic process. Our analysis demonstrates that under those biological constraints oxidation of alkanes to acetate and $H_2$, linked to syntrophic acetate oxidation and methanogenesis from $CO_2$ reduction offers a greater window of opportunity than complete oxidation of alkanes to $H_2$ and $CO_2$ linked to methanogenic $CO_2$ reduction, and hence is the process most likely to occur.

# INTRODUCTION

The largest deposits of petroleum on Earth are not, as conventionally assumed, in the Middle East. The vast Saudi Arabian and Kuwaiti oilfields of Ghawar ($2.6 \times 10^{11}$ barrels (bbl) in place) and Burgan ($7.0 \times 10^{10}$ bbl in place) are dwarfed by the trillion bbl deposits of western Canada (Athabasca tar sands; $1.7 \times 10^{12}$ bbl) and Venezuela (Orinoco heavy oil belt; $1.2 \times 10^{12}$ bbl). These so-called super-giant heavy oil fields are the result of biodegradation of the lighter, more readily produced and valuable oil fractions over geological time. Biodegraded oil fields are more difficult to produce and the oils more difficult to refine than oil from conventional fields and are thus less economically attractive. On the basis of known mechanisms of hydrocarbon degradation, conventional wisdom among petroleum geologists has for some time, been that biodegradation in oilfields was driven by oxygen delivered to petroleum reservoirs in meteoric water. This paradigm has been questioned in light of the discovery of a range of bacteria capable of coupling the oxidation of aliphatic or aromatic hydrocarbons to the reduction of nitrate, iron and sulphate (Widdel and Rabus, 2001) and microbial consortia capable of linking aliphatic hydrocarbon oxidation to methane generation (Zengler et al., 1999;Anderson and Lovley, 2000; Townsend et al., 2003).

Evidence is emerging to support the notion that in-reservoir petroleum biodegradation is caused by anaerobic hydrocarbon degrading bacteria. Reduced naphthoic acids, metabolites characteristic of anaerobic hydrocarbon degradation have been detected in biodegraded petroleum reservoirs, but not in non-degraded reservoirs (Aitken et al., 2004). Geochemical and isotopic evidence also suggests that in many cases the end product of hydrocarbon degradation in petroleum reservoirs is methane (Scott et al., 1994; Larter et al., 1999; Sweeney and Taylor, 1999;Pallasser, 2000; Boreham et al., 2001; Masterson et al., 2001). Compositional gradients in oil columns towards the underlying water leg in biodegraded petroleum reservoirs suggests that the oil water transition zone is the primary site of biodegradation in petroleum reservoirs (Head et al., 2003; Larter et al., 2003).

On the basis of these and other data, a new conceptual model of in-reservoir petroleum biodegradation has been developed (Head et al., 2003). In this model, anaerobic degradation of petroleum occurs most actively at the oil water transition zone. Electron donor, mainly hydrocarbons, is delivered to the oil water transition zone by diffusion from the oil column, with inorganic nutrients such as ammonium ions provided from the water leg (Head et al., 2003;Manning and Hutcheon, 2004). This is consistent with reports from other deep subsurface environments that microbial activity is stimulated at geochemical interfaces (Parkes et al., 2005). When the water leg contains low levels of sulphate, hydrocarbon degradation is driven by methanogenesis; indeed, many biodegraded petroleum reservoirs contain isotopically light methane indicative of a mixed secondary biogenic and thermogenic source (Scott et al., 1994; Larter et al., 1999; Sweeney and Taylor, 1999;Pallasser, 2000; Boreham et al., 2001; Masterson et al., 2001;Head et al., 2003).

The significance of methanogenic crude oil degradation in petroleum reservoirs goes beyond its potential role in the biodegradation of petroleum reservoirs; it may ultimately be crucial for processes that can enhance the recovery of residual oil. Typically, over 60% of the oil in place in a petroleum reservoir remains unextractable following standard production procedures and the possibility that methanogenic degradation of this residual oil can re-pressurize a petroleum reservoir, has some potential for enhancing oil recovery. Furthermore, the volumetrics of gas recovery are far better than for oil (typically 70% of gas in place can be recovered) and methanogenic conversion of non-recoverable residual hydrocarbons to recoverable gas may be an economically viable way of extending the operational life of petroleum reservoirs (Parkes, 1999; Larter *et al.*, 1999; Head *et al.*, 2003). In addition, methanogenic hydrocarbon degradation may be a significant process in the attenuation of contaminated anoxic sediments and aquifers (Weiner and Lovley, 1998; Anderson and Lovley, 2000; Bekins *et al.*, 2005).

Because of the potential importance of methanogenic crude oil biodegradation and our limited knowledge of the organisms and mechanisms involved, it is important that we learn more about what governs the microbial conversion of oil to methane.

Quantitatively, the most important component of crude oil is the saturated hydrocarbon fraction and little is known about the methanogenic degradation of long-chain aliphatic hydrocarbons; only three reports in the literature provide strong evidence of methanogenic degradation of aliphatic hydrocarbons or crude oil. Zengler *et al.* (1999) report degradation of pure hexadecane by an enrichment culture; Anderson and Lovley (2000) documented rapid mineralization of [14]C-labelled hexadecane in sediments from a crude oil-contaminated aquifer; and Townsend *et al.* (2003) observed methanogenic transformation of crude oil in sediments from a gas condensate contaminated aquifer.

In this paper, we evaluate the thermodynamics of five possible routes of methanogenic hydrocarbon degradation, viz (with hexadecane as example):

- complete oxidation of alkanes to $H_2$ and $CO_2$, linked to methanogenesis from $CO_2$ reduction:

$$4C_{16}H_{34} + 128H_2O \rightarrow 64CO_2 + 196H_2 \quad \text{(reaction 1)}$$

$$196H_2 + 49CO_2 \rightarrow 49CH_4 + 98H_2O \quad \text{(reaction 2)}$$

$$\text{sum } 4C_{16}H_{34} + 30H_2O \rightarrow 15CO_2 + 49CH_4 \quad \text{(reaction 3)}$$

- oxidation of alkanes to acetate and $H_2$, linked to acetoclastic methanogenesis and $CO_2$ reduction:

$$4C_{16}H_{34} + 64H_2O \rightarrow 32CH_3COO^- + 32H^+ + 68H_2$$
$$\text{(reaction 4)}$$

$$32CH_3COO^- + 32H^+ \rightarrow 32CO_2 + 32CH_4 \quad \text{(reaction 5)}$$

$$68H_2 + 17CO_2 \rightarrow 17CH_4 + 34H_2O \quad \text{(reaction 6)}$$

$$\text{sum } 4C_{16}H_{34} + 30H_2O \rightarrow 15CO_2 + 49CH_4$$

- oxidation of alkanes to acetate and $H_2$, linked to syntrophic acetate oxidation and methanogenesis from $CO_2$ reduction:

$$4C_{16}H_{34} + 64H_2O \rightarrow 32CH_3COO^- + 32H^+ + 68H_2$$

$$32CH_3COO^- + 32H^+ + 64H_2O \rightarrow 64CO_2 + 128H_2 \quad \text{(reaction 7)}$$

$$196H_2 + 49CO_2 \rightarrow 49CH_4 + 98H_2O$$

$$\text{sum } 4C_{16}H_{34} + 30H_2O \rightarrow 15CO_2 + 49CH_4$$

- oxidation of alkanes to acetate alone, linked to acetoclastic methanogenesis:

$$4C_{16}H_{34} + 30H_2O + 34CO_2 \rightarrow 49CH_3COO^- + 49H^+$$
$$\text{(reaction 8)}$$

$$49CH_3COO^- + 49H^+ \rightarrow 49CO_2 + 49CH_4 \quad \text{(reaction 9)}$$

$$\text{sum } 4C_{16}H_{34} + 30H_2O \rightarrow 15CO_2 + 49CH_4$$

And

- oxidation of alkanes to acetate alone, linked to syntrophic acetate oxidation and methanogenesis from $CO_2$ reduction:

$$4C_{16}H_{34} + 30H_2O + 34CO_2 \quad \rightarrow 49CH_3COO^- + 49H^+$$
(reaction 10)

$$49CH_3COO^- + 49H^+ + 98H_2O \rightarrow 98CO_2 + 196H_2$$
(reaction 11)

$$196H_2 + 49CO_2 \quad \rightarrow 49CH_4 + 98H_2O$$
(reaction 12)

$$\text{sum } 4C_{16}H_{34} + 30H_2O \quad \rightarrow 15CO_2 + 49CH_4$$

The effects of temperature, pH, acetate and $H_2$ concentration on each of these processes is determined and conditions under which each process is likely to be most favourable are identified and related to conditions typical of petroleum reservoirs. We consider only the case of alkanes with even chain length. It is likely that similar patterns will be observed with odd chain alkanes, which will generate propionate in addition to acetate and $H_2$. This would require incorporation of syntrophic propionate oxidation as an intermediate reaction. Any effect of propionate will decrease with increasing alkane chain length.

# METHODS

Gibbs free energy calculations were made after Thauer *et al.* (1977) and Amend and Shock (2001).

Temperature corrections for $G^o$ were made with the Gibbs–Helmholtz equation according to with T in K; $T_{ref} = 298.15$ K.

$$\Delta G^o_{Tact} = \Delta G^o_{Tref} \cdot (T_{act}/T_{ref}) + \Delta H^o_{Tref} \cdot (T_{ref} - T_{act})/T_{ref}$$

Gibbs free energies and enthalpies of formation data for alkanes in the liquid state were taken from (Helgeson *et al.*, 1998). For all other compounds the data were taken from Hanselmann (Hanselmann, 1991), with acetate in the aqueous phase and methane, hydrogen and carbon dioxide in the gaseous phase at partial pressures of 1 atm. Calculations were made for neutrality rather than for pH=7, with neutrality defined as the pH where activities of $H^+$ and $OH^-$ are equal. The pH representing neutrality varies with temperature (Harned and Owen, 1943).

# Sample Calculations

Table 1 lists change in Gibbs free energy values for various reactions of importance to anaerobic hexadecane degradation. The values shown are $G^\circ$. From these data the threshold concentrations of products and reactants, which result in $G'<0$ were calculated as follows. The example given is for the threshold $H_2$ concentration for hydrogenotrophic methanogenesis.

$$4H_2 + CO_2 \rightarrow CH_4 + 2H_2O;$$
$$\Delta G^{o'} = -130.7 kJ/molCH_4$$

**Table 1:** Stoichiometry and change in Gibbs free energy values for reactions potentially involved in methanogenic hexadecane degradation

| Substrates | Products | $\Delta G^{o'}$ | |
|---|---|---|---|
| | | *kJ/reaction* | *kJ/mol* |
| *Hexadecanedegradation* | | | |
| $4C_{16}H_{34}+128H_2O$ | $64CO_2+196H_2$ | 4922.1 | 1230.5[a] |
| $4C_{16}H_{34}+64H_2O$ | $32CH_3COO^-$ $+32H^++68H_2$ | 1883.1 | 470.8[a] |
| $4C_{16}H_{34}+30H_2O+34CO_2$ | $49CH_3COO^-+49H^+$ | 268.6 | 67.2[a] |
| $4C_{16}H_{34}+30H_2O$ | $15CO_2+49CH_4$ | -1487.1 | -371.8[a] |
| | | | |
| *Conversionofpotentialintermediates* | | | |
| $CH_3COO^-+H^++2H_2O$ | $2CO_2+4H_2$ | 94.9 | 94.9[b] |
| $CH_3COO^-+H^+$ | $CO_2+CH_4$ | -35.8 | -35.8[c] |
| $4H_2+CO_2$ | $CH_4+2H_2O$ | -130.7 | -130.7[c] |

[a]Per mol hexadecane.

[b]Per mol acetate.

[c]Per mol methane.

Hence (for example Thauer et al., 1977): $\Delta G' = -130.7 + RT \ln ([CH_4]/[CO_2].[H_2]^4)$

(note that in biological systems $\ln[H_2O]$ is assumed to be 0)

Therefore, under otherwise standard conditions: $\Delta G' = -130.7 - 5.71 \log [H_2]^4$ (where $5.71 \log x$ equals $R.T_{298.15} \ln x$).

Since the threshold value is the value where $\Delta G' = 0$ it follows that

$$- 130.7 - 5.71 \log[H_2]^4 = 0$$
$$\text{Thus } 4 \log[H_2] = 130.7/ - 5.71, \text{ and hence}$$
$$[H_2] = 10^{(-130.7/22.84)} = 10^{-5.72}$$
$$\text{Thus } [H_2]_{crit} = 1.89 \times 10^{-6} \text{ atm.}$$

# RESULTS

We have examined the effect of alkane chain length on the free energy yield of methanogenic alkane degradation. Hexadecane was chosen as an exemplar to determine the effect of temperature, $H_2$ concentration, acetate concentration and pH on the thermodynamics of alkane degradation via routes (i) to (v) outlined above. By determining threshold conditions at which the different component reactions for each route become exergonic we have defined 'windows of opportunity' under which different individual reactions can be linked, and thus what conditions are permissive for each of the five routes. The windows of opportunity determined, are discussed in the context of empirical data on crude oil degradation in petroleum reservoirs and used to assess the relative importance of thermodynamics relative to biological factors in determining the routes of methanogenic crude oil degradation, observed in nature.

# Effect of Chain Length on the Thermodynamics of Anaerobic Alkane Degradation

The effect of chain length on the free energy yield of methanogenic degradation of alkanes was investigated to evaluate if inferences made using a single model alkane would be generally applicable across all alkanes.

Thermodynamic calculations for all alkanes in the range $C_8$–$C_{80}$, demonstrated that conversion to methane was exergonic. $\Delta G°$ varied between −185 and −1870 kJ per mol for $C_8$ to $C_{80}$ alkanes respectively (Figure 1). In all cases, the free energy yield increased with increasing temperature and the increase in energy yield per degree Kelvin was greater for longer chain hydrocarbons (−0.93 kJ mol$^{-1}$ K$^{-1}$ for $C_8$ and −10.0 kJ mol$^{-1}$ K$^{-1}$ for $C_{80}$). Similar trends were observed for complete oxidation of alkanes to $H_2$ and $CO_2$ and incomplete oxidation of alkanes to $H_2$ and acetate (reactions 1 and 4). In contrast, oxidation of alkanes to acetate alone (reaction 8) became less energetically favourable with increasing temperature, especially for longer chain alkanes (+0.73 kJ mol$^{-1}$ K$^{-1}$ for $C_8$ and +6.0 kJ mol$^{-1}$ K$^{-1}$ for $C_{80}$: see Supplementary Information for data). When normalized for the number of carbon atoms, there is a slightly greater energy yield per carbon from longer chain length alkanes (for $C_8$ $\Delta G°$=−23.1 kJ per mol C, for $C_{80}$ $\Delta G°$=−23.4 kJ per mol C).

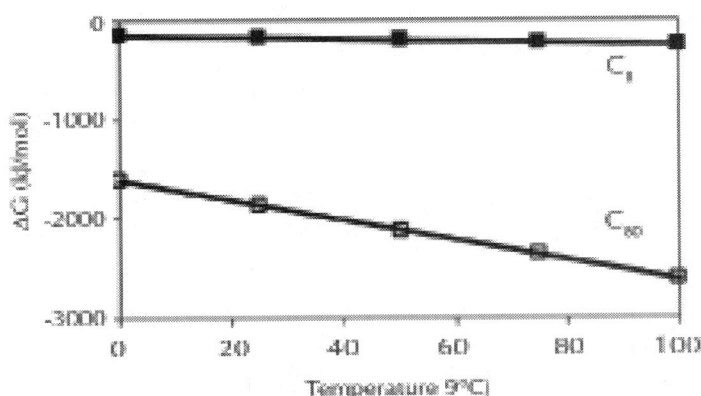

**Figure 1:** Effect of temperature on change in Gibbs free energy for methanogenesis from alkanes; closed symbols $C_8$, open symbols $C_{80}$.

These results relate to standard conditions, where the initial oxidation reactions (reactions 1, 4 and 8) are all endergonic. When the effect of $H_2$ on the complete oxidation pathway is taken into account, it is clear that $H_2$ concentration has a more marked effect on long chain alkanes than short chain alkanes (Figure 2). However, it is interesting to note that the threshold $H_2$ concentration at which alkane oxidation to $H_2$ and $CO_2$ becomes energetically favourable under otherwise standard conditions, increases merely from $3.7 \times 10^{-5}$ atm for $n$-octane ($C_8$) to $4.3 \times 10^{-5}$ atm for $n$-octacontane ($C_{80}$), that is, is essentially independent of chain length (Figure 2).

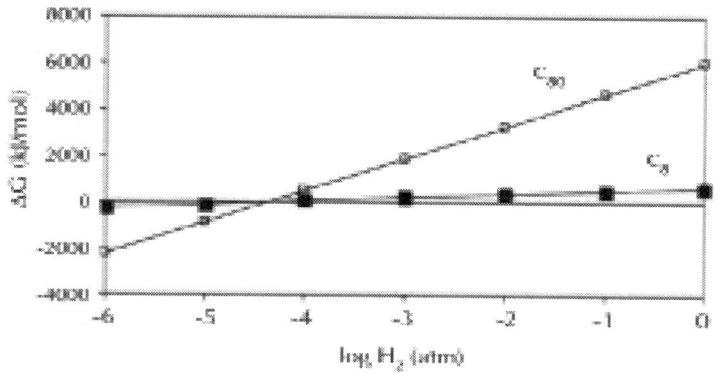

**Figure 2:** Effect of hydrogen partial pressure on the change in Gibbs free energy for oxidation of alkanes to $H_2$ plus $CO_2$; closed symbols $C_8$, open symbols $C_{80}$.

Overall the same trends are observed, independent of alkane chain length and for simplicity all subsequent discussion is restricted to analysis of hexadecane oxidation, which we have used as an example to illustrate the effect of hydrogen concentration, acetate concentration, temperature and pH on different pathways of methanogenic alkane degradation.

## Complete Oxidation of Hexadecane to $H_2$ and $Co_2$ Linked to Methanogenic $Co_2$ Reduction

While hexadecane oxidation to $H_2$ and $CO_2$ becomes more thermodynamically favourable with increasing temperature, methanogenesis by $CO_2$ reduction becomes less thermodynamically favourable with increasing temperature (Figure 3). However, because the energy yield from hexadecane oxidation increases more steeply than the reduction in energy yield for the methanogenic reaction, the net result is an increase in the overall energy yield from methanogenic hexadecane degradation with increasing temperature (Figure 3).

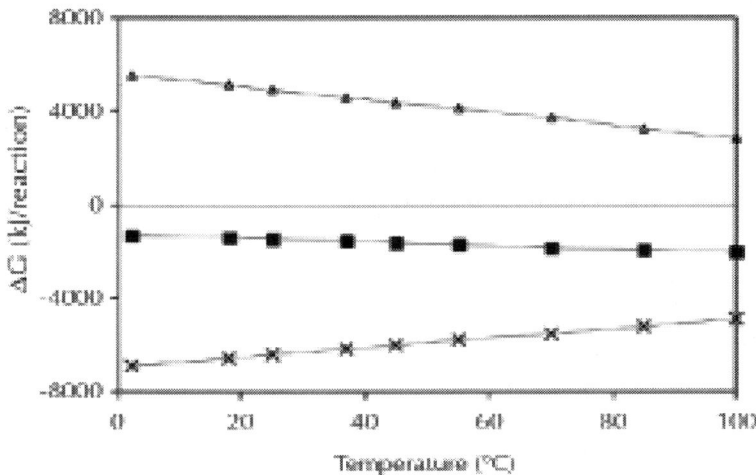

**Figure 3:** Effect of temperature on the change in Gibbs free energy for complete oxidation of hexadecane to $H_2$ and $CO_2$ (open symbols), for stoichiometric methanogenesis of the hydrogen produced (+) and for the sum of the aforementioned reactions, that is, methanogenic degradation of hexadecane (closed symbols). Reactions considered: $4C_{16}H_{34}+128H_2O \rightarrow 196H_2+64CO_2$; $196H_2+49CO_2 \rightarrow 49CH_4+98H_2O$; $4C_{16}H_{34}+30H_2O \rightarrow 15CO_2+49CH_4$.

When the $H_2$ concentration is taken into account, the energy yield from complete oxidation of hexadecane decreases with increasing $H_2$ concentration (Figure 4) whereas methanogenic $CO_2$ reduction becomes more exergonic with increasing $H_2$ concentration. Provided the overall reaction is exergonic, there is a 'window of opportunity' defined by the $H_2$ concentrations where both processes are exergonic (Figure 4). It is particularly interesting to note that this window of opportunity increases with increasing temperature and thus this pathway of methanogenic hexadecane degradation becomes less constrained with increasing temperature (that is the window of opportunity becomes larger. Note that the $y$ axis is on a log scale (Figure 5)).

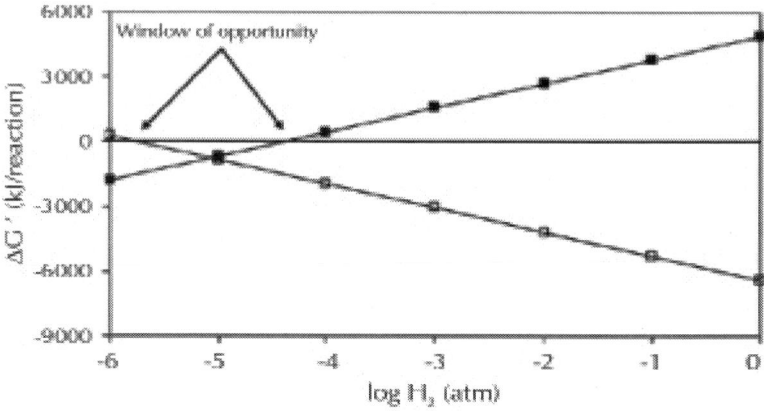

**Figure 4:** Effect of hydrogen partial pressure on the change in Gibbs free energy for oxidation of hexadecane to $H_2$ and $CO_2$ (open symbols) and for stoichiometric methanogenesis of the hydrogen produced. The arrows delineate the 'window of opportunity' where both reactions are exergonic. Reactions considered: $4C_{16}H_{34}+128H_2O \rightarrow 196H_2+64CO_2$; $196H_2+49CO_2 \rightarrow 49CH_4+98H_2O$.

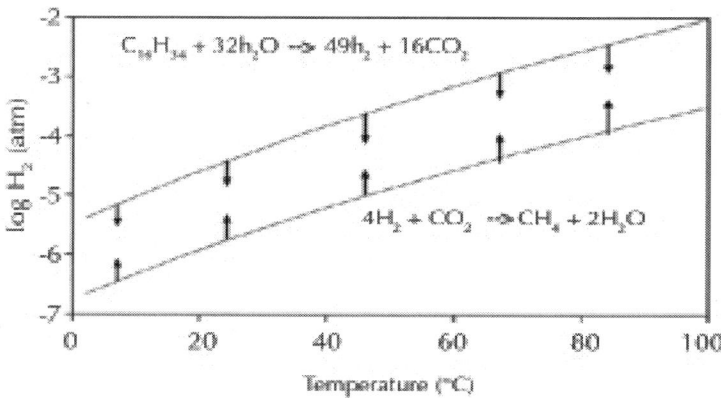

**Figure 5:** Effect of temperature on the range of $H_2$ partial pressures where both hexadecane oxidation and methanogenesis from $H_2/CO_2$ are exergonic. The lines represent the threshold at which the free energy change for each process is equal to zero and the arrows indicate conditions under which the processes become increasingly exergonic.

# Incomplete Oxidation of Hexadecane to Acetate and $H_2$ Linked to Acetoclastic Methanogenesis and Methanogenic $Co_2$ Reduction

As with complete oxidation of hexadecane, the free energy yield of incomplete oxidation of hexadecane to acetate and $H_2$ is highly dependent on $H_2$ concentration (Figure 6). The range of $H_2$ concentrations where oxidation of hexadecane to acetate and $H_2$ is exergonic is also dependent on the acetate concentration (Figure 7). As acetate concentration decreases, the range of $H_2$ concentrations that are permissive for incomplete oxidation of hexadecane linked to methanogenic $CO_2$ reduction increases. Furthermore, the higher the temperature the more permissive incomplete hexadecane oxidation becomes with respect to $H_2$ concentration (Figure 7). The effect of acetate concentration on incomplete oxidation of hexadecane is therefore considerable, but is dependent upon $H_2$ concentration.

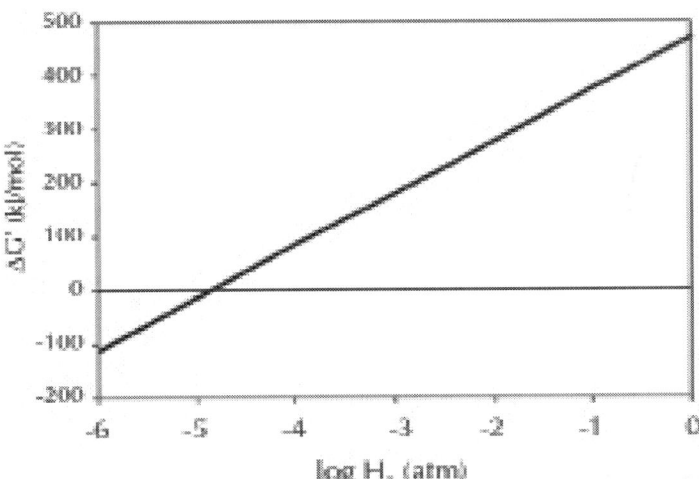

**Figure 6:** Effect of hydrogen partial pressure on the change in Gibbs free energy for incomplete oxidation of hexadecane to acetate and $H_2$, under otherwise standard conditions, that is acetate 1 m, hexadecane as liquid and pH=7.

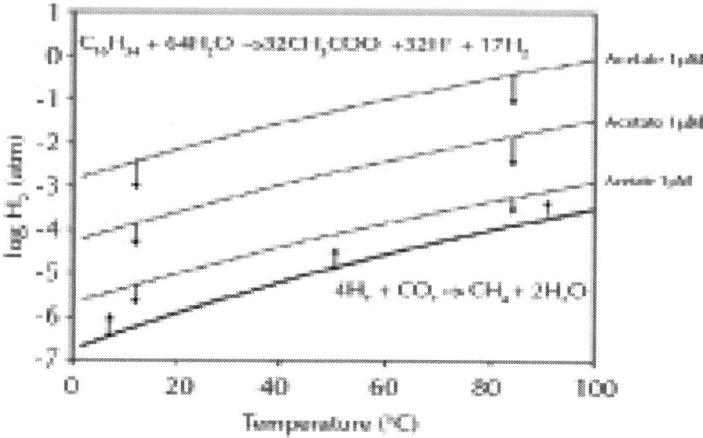

**Figure 7:** Effect of temperature on the range of $H_2$ partial pressures where both incomplete oxidation of hexadecane to acetate and $H_2$ and methanogenic $CO_2$ reduction are exergonic. The thin lines demarcate the hydrogen thresholds where the free energy yield equals zero at acetate concentrations of 1 μm, 1 mm and 1 m respectively. The thick line gives the threshold for methanogenic $CO_2$ reduction. The arrows indicate conditions under which the processes become increasingly exergonic.

## Oxidation of Hexadecane to Acetate and $H_2$ Linked to Syntrophic Acetate Oxidation and Methanogenesis from $Co_2$ Reduction

A variant of incomplete hexadecane oxidation linked to acetoclastic methanogenesis and methanogenic $CO_2$ reduction could involve syntrophic acetate oxidation linked exclusively to methanogenic $CO_2$ reduction, with no involvement of acetoclastic methanogenesis. Crude oil has been shown to have an adverse effect on acetoclastic methanogenesis (Warren et al., 2004); under such circumstances it is possible that syntrophic acetate oxidation might permit complete conversion of hexadecane to methane and $CO_2$ via methanogenic $CO_2$ reduction. Interestingly, the range of $H_2$ concentrations that permit hexadecane oxidation linked to methanogenic $CO_2$ reduction is least for incomplete oxidation and greatest for syntrophic acetate oxidation, which is exergonic at $H_2$ partial pressures slightly greater than required to give a negative G for complete oxidation of hexadecane to $H_2$ and $CO_2$ (Figure 8). However, as acetate concentrations decrease (less than 0.1 macetate), syntrophic acetate oxidizers require lower $H_2$ concentrations than either complete or incomplete hexadecane oxidizers (Figure 9; sector II and III compared to sector IV) and the feasibility of this route is therefore dictated by the sensitivity of syntrophic acetate oxidation to $H_2$ concentration.

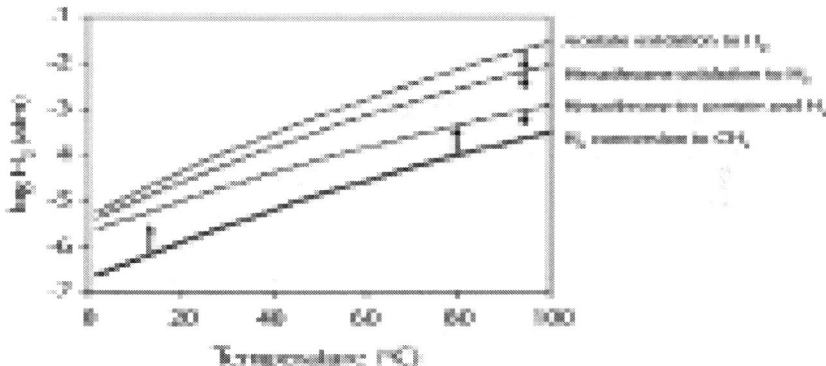

**Figure 8:** Effect of temperature on the range of $H_2$ partial pressures where complete oxidation of acetate to $H_2$, complete oxidation of hexadecane to $H_2$ and $CO_2$ and incomplete oxidation of hexadecane to acetate and $H_2$ are exergonic and can be coupled to methanogenic $CO_2$ reduction. The thin lines demarcate the hydrogen thresholds for acetate oxidation and for hexadecane oxidation; the thick line gives the threshold for methanogenic $CO_2$ reduction. Standard states are: acetate 1 m, hexadecane as liquid and $CO_2$ and $CH_4$ at 1 atm.

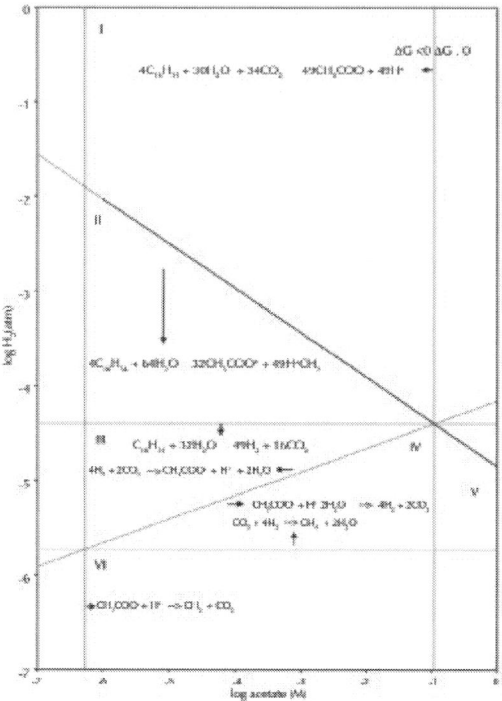

**Figure 9:** Hydrogen and acetate as thermodynamic constraints on methanogenic hexadecane degradation.

# Oxidation of Hexadecane to Acetate Alone, Linked to Acetoclastic Methanogenesis

A further route for the primary oxidation of alkanes is their conversion to acetate alone, which may then be linked to acetoclastic methanogenesis. This pathway becomes thermodynamically less favourable with increasing temperature (Figure 10). Like incomplete hexadecane oxidation, this mechanism of primary oxidation of hexadecane, is dependent on the acetate concentration. Under otherwise standard conditions, this route of hexadecane oxidation becomes exergonic at concentrations less than 0.1 m acetate (Figure 9). This contrasts with incomplete hexadecane oxidation which is exergonic at higher acetate concentrations provided that the $H_2$ partial pressure is maintained at less than $1.4 \times 10^{-5}$ atm (Figure 9, sector V; see also section 3.4).

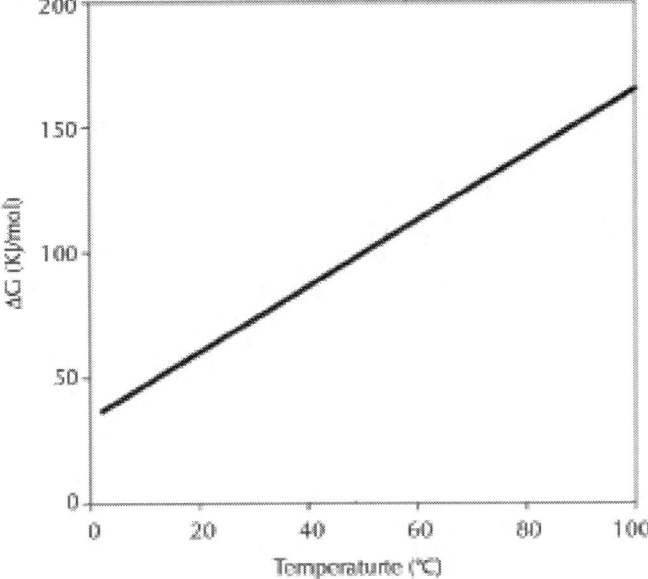

**Figure 10:** Effect of temperature on change in Gibbs free energy for oxidation of hexadecane to acetate alone.

## Oxidation of Hexadecane to Acetate Alone Linked to Syntrophic Acetate Oxidation and Methanogenic Co₂ Reduction

A variation on complete acetate oxidation linked to methanogenesis involves syntrophic acetate oxidation and methanogenic $CO_2$ reduction as the terminal reaction. This metabolic route has the effect of reducing the domain under which conversion of hexadecane to methane is feasible (Sector I through VI in Figure 9, for hexadecane conversion to acetate linked to acetoclastic methanogenesis, compared to Sector I through IV inFigure 9 if syntrophic acetate oxidation was involved). This is because acetoclastic methanogenesis is independent of $H_2$ concentration, but methanogenic $CO_2$ reduction is prevented at low $H_2$ concentrations, and in addition oxidation of hexadecane to acetate alone is prevented at acetate concentrations greater than 0.1 m (Figure 9).

# Effect of pH on Incomplete Oxidation and Acetogenic Oxidation of Hexadecane

Both incomplete oxidation of hexadecane and conversion of hexadecane to acetate are pH dependent. Both processes become increasingly energetically favourable with increasing pH. In line with the stoichiometry of acetate production, the effect is more pronounced for complete conversion of hexadecane to acetate (Figure 11). Higher pH values also increase the range of acetate concentrations where the reaction remains exergonic (Figure 12).

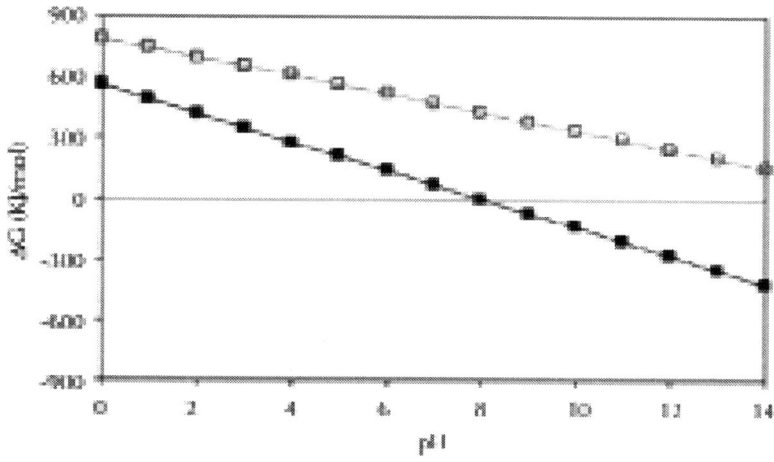

**Figure 11:** Effect of pH on change in Gibbs free energy for complete oxidation of hexadecane to acetate (closed symbols) and for incomplete oxidation to $H_2$ and acetate (open symbols).

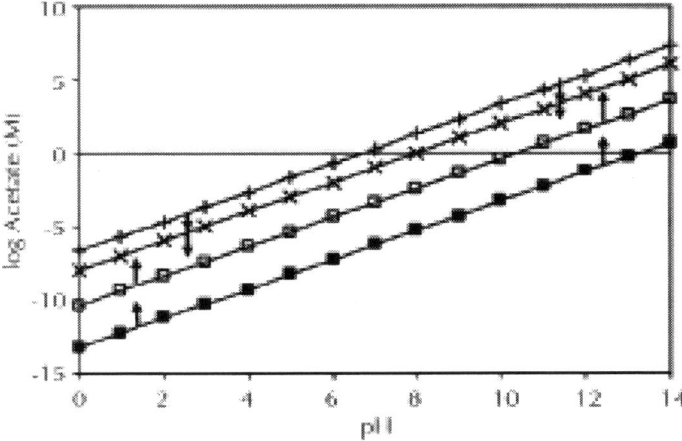

**Figure 12:** Effect of pH on the range of acetate concentrations where hexadecane conversion to acetate or acetate plus hydrogen are exergonic and can be coupled to methanogenesis from acetate at a hydrogen partial pressure of $10^{-5}$ atm; symbols: +, incomplete oxidation of hexadecane to acetate and $H_2$; x, oxidation of hexadecane to acetate; open squares, oxidation of acetate to $H_2$ and $CO_2$; closed squares, conversion of acetate to $CH_4$. The arrows indicate conditions under which the processes become increasingly exergonic.

# DISCUSSION

## The Methanogenic Alkane Degradation Landscape

Microbial degradation of hydrocarbons under anoxic conditions is of global significance. It is responsible for the in-reservoir degradation of oil that has lead to vast heavy oil and tar sand deposits that constitute the bulk of the world's petroleum inventory (Roadifer, 1987; Head et al., 2003; Aitken et al., 2004). Despite the importance of the process, relatively little is known about the factors which dictate the occurrence of in-reservoir biodegradation. The geothermal history of petroleum basins has been shown to exert a broad control on the occurrence of biodegraded petroleum reservoirs and has led to the concept of palaeopasteurization or palaeosterilization, which prevents the biodegradation of crude oil in subterranean formations (Wilhelms *et al.*, 2001). The limited availability of nutrients in aquifers associated with petroleum reservoirs and geological constraints on the physical interaction of water and oil legs have also been proposed as factors which may limit in-reservoir oil biodegradation (Head *et al.*, 2003; Larter *et al.*, 2006). As we have seen, there are several possible routes that lead to the conversion of crude oil hydrocarbons to methane, and the interplay between physical and chemical conditions in the reservoir and reservoir microbiology likely have a role in controlling in-reservoir oil biodegradation. In addition to these broad scale geological controls, thermodynamic factors may also have a bearing on crude oil biodegradation in petroleum reservoirs.

Thermodynamic calculations demonstrate that methanogenic alkane degradation becomes more energetically favourable at higher temperatures and although the three potential initial oxidation reactions examined here are all endergonic under standard conditions, at low concentrations of $H_2$ and/or acetate they are all thermodynamically feasible. Acetate concentrations ranging from less than 10 µm to over 17 mm have been measured in oil-field formation waters (Barth and Riis, 1992) with the majority (85 of 121 samples analyzed) having acetate concentrations below 5 mm. Such concentrations are well within the permissive range for the hydrocarbon oxidation reactions investigated here (Figure 9). There are very few data in the literature on hydrogen gas concentrations in petroleum reservoirs. Hydrogen however, seems to be present below detection limits in most oilfield waters examined and is unlikely to represent more than 1 mol% (equivalent to $10^{-2}$ atm) in associated gases (Hill *et al.*, 2007). This again is compatible with the hydrocarbon oxidation pathways examined here (Figure 9). However, it should be borne in mind that the gas data were obtained from well-head samples and may not reflect the true concentrations present in deep petroleum reservoirs.

To evaluate how the different initial oxidation processes can be linked to conventional methanogenic pathways we have introduced the concept of a 'window of opportunity', which defines the conditions under which both the initial oxidation process and the terminal methanogenic pathways are energetically favourable. This demonstrates that as temperature increases, so does the window of opportunity with respect to $H_2$ and acetate. Clearly this is only part of the story as it is well known that at temperatures in excess of 80–90 °C, in-reservoir petroleum biodegradation apparently ceases (Connan, 1984; Head et al., 2003), illustrating that above these temperatures biological factors are more important than thermodynamic factors in controlling methanogenic hydrocarbon degradation. The windows of opportunity with respect to acetate and $H_2$, the central intermediates in methanogenic alkane degradation, have been summarized for the range of processes that are feasibly involved in methanogenic alkane degradation. Figure 9 represents the situation at 25 °C and pH 7; however, the activity domains identified from this analysis remain similar for other temperatures and pH values, with the proviso that they are shifted towards higher $H_2$ and lower acetate concentrations at higher temperatures (see Supplementary Information). The value of this analysis is that it identifies clear zones where different methanogenic alkane degradation pathways can occur. For example the window of opportunity for linking complete conversion of hexadecane to acetate with acetoclastic methanogenesis (domain I to IV and VI in Figure 9) is much larger than the equivalent window for linking incomplete oxidation of hexadecane to both acetoclastic methanogenesis and methanogenic $CO_2$ reduction (domain II, III, IV and VI in Figure 9), or complete oxidation of hexadecane to methanogenic $CO_2$ reduction (domain III, IV and V in Figure 9). This might suggest that acetoclastic methanogenesis might predominate in methanogenic oil-degrading systems. However, several lines of evidence suggest that methanogenic $CO_2$ reduction, may be more prevalent in petroleum systems. The majority of methanogens identified from oil field waters are $CO_2$-reducing methanogens, and acetoclastic methanogens are apparently rare (Magot et al., 2000; Orphan et al., 2000, 2003; Grabowski et al., 2005). Furthermore, experimental measurements of methanogenic pathways in oil field waters also suggest that $CO_2$ reduction to methane may be more important than acetoclastic methanogenesis in petroleum reservoirs. A synthesis of data from 138 measurements made across 6 different oilfields (Nazina et al., 1995a, 1995b; Rozanova et al., 1995, 2001; Bonch-Osmolovskaya et al., 2003) indicates that in 67% of the cases, the predominant methanogenic pathway was $CO_2$ reduction, whereas in 33% acetoclastic methanogenesis predominated. This is supported by field data which demonstrate that acetate is present at high concentrations in some oil field waters (Barth and Riis, 1992), and by modelling of gas isotope composition during in-reservoir biodegradation and data from methanogenic oil-degrading microcosms (Jones et al., 2008 accepted for publication). Studies of methanogenesis in a crude oil contaminated aquifer also suggest that crude oil suppresses acetoclastic methanogenesis (Warren et al., 2004). Interestingly predominance of acetoclastic methanogens has been observed in hydrocarbon contaminated aquifers (Dojka et al., 1998; Struchtemeyer et al., 2005). However, in these instances, refined petroleum products (jet fuel) and a gas condensate were the contaminants, not crude oil. Taken together these suggest that there are biological controls that override thermodynamic considerations and lead to a subordinate role for processes of crude oil alkane degradation that require a major role for acetoclastic methanogenesis. The thermodynamic analysis; however, demonstrates that there is a wide range of conditions under which alkane degradation can be supported by methanogenic $CO_2$ reduction linked either to the complete oxidation of alkanes (domain III to V), or incomplete oxidation of alkanes where the acetate generated can be oxidized to $H_2$ and $CO_2$ by syntrophic acetate oxidizing organisms (domain II to V in Figure 9). The window of opportunity, which is permissive for the pathway involving incomplete alkane oxidation coupled to syntrophic acetate oxidation and methanogenic $CO_2$ reduction is larger than the corresponding window for complete alkane oxidation coupled to methanogenic $CO_2$ reduction. Furthermore, if one considers the line relating $\Delta G$ to hydrogen concentration for complete and incomplete alkane oxidation, it is apparent that incomplete oxidation is more favourable at higher hydrogen concentrations and the slope is much steeper for complete oxidation (see Supplementary Figure S6). The second of these observations indicates that complete oxidation of alkanes is likely be more sensitive to small fluctuations in hydrogen concentrations around the point that the process shifts from being exergonic to endergonic. Thus under conditions where acetoclastic methanogenesis is inhibited one would expect a pathway via incomplete oxidation of alkanes linked to syntrophic acetate oxidation and methanogenic $CO_2$ reduction, rather than a pathway via complete oxidation to $H_2/CO_2$ linked to methanogenic $CO_2$ reduction. Interestingly, putative thermophilic acetate-oxidizing bacteria have been detected in a petroleum reservoir (Nazina et al., 2006). Coupled with the inability to detect acetoclastic methanogens in formation waters or enrichment cultures, even when high rates of methanogenesis from acetate were measured, this has led to the suggestion that syntrophic acetate oxidation might be important for driving methanogenesis in a high temperature oil reservoir (Nazina *et al.*, 2006). Therefore it is feasible that the principal pathways leading to methanogenic oil degradation in petroleum reservoirs involve either complete oxidation of hydrocarbons to $H_2$ and $CO_2$ followed by methanogenic $CO_2$ reduction or incomplete hydrocarbon oxidation coupled with syntrophic acetate oxidation and methanogenic $CO_2$ reduction with the latter of these likely to occur more widely.

# ACKNOWLEDGEMENTS

JD and IMH acknowledge funding from the European Commission, which supported this work through ECOSERV, a Marie Curie Excellence Grant (EXT 023469) and the Natural Environment Research Council (Grant No NE/E01657X/1). SRL acknowledges support from Alberta Ingenuity, NSERC and CFI.

# REFERENCES

1.   Aitken CM, Jones DM, Larter SR. (2004). Anaerobic hydrocarbon biodegradation in deep subsurface oil reservoirs.*Nature* 431: 291–294

2.   Amend JP, Shock EL. (2001). Energetics of overall metabolic reactions of thermophilic and hyperthermophilic archaea and bacteria. *FEMS Microbiol Rev* 25: 175–243.

3.   Anderson RT, Lovley DR. (2000). Hexadecane decay by methanogenesis. *Nature* 404: 722–723

4.   Barth T, Riis M. (1992). Interactions between organic acid anions in formation waters and reservoir mineral phases. *Org Geochem* 19: 455–482.

5.   Bekins BA, Hostettler FD, Herkelrath WN, Delin GN, Warren E, Essaid HI. (2005). Progression of methanogenic degradation of crude oil in the subsurface. *Environ Geosci* 12: 139–152

6.   Bonch-Osmolovskaya EA, Miroshnichenko ML, Lebedinsky AV, Chernyh NA, Nazina TN, Ivoilov VS *et al.* (2003). Radioisotopic, culture-based and oligonucleotide microchip analyses of

thermophilic microbial communities in a continental high-temperature petroleum reservoir. *Appl Environ Microbiol* 69: 6143–6151.

7.    Boreham CJ, Hope JM, Hartung-Kagi B. (2001). Understanding source, distribution and preservation of Australian natural gas: a geochemical perspective. *APPEA J* 41: 523–547.

8.    Connan J. (1984). Biodegradation of crude oils in reservoirs. In: Brooks J, Welte DH (eds). *Advances in Petroleum Geochemistry*, vol. 1. Academic Press: London, pp 299–335.

9.    Dojka MA, Hugenholtz P, Haack SK, Pace NR. (1998). Microbial diversity in a hydrocarbon- and chlorinated-solvent-contaminated aquifer undergoing intrinsic bioremediation.*Appl Environ Microbiol* 64: 3869–3877

10.   Grabowski A, Nercessian O, Fayolle F, Blanchet D, Jeanthon C. (2005). Microbial diversity in production waters of a low-temperature biodegraded oil reservoir. *FEMS Microbiol Ecol*54: 427–443

11.   Hanselmann KW. (1991). Microbial energetics applied to waste repositories. *Experientia* 47: 645–687.

12.   Harned HS, Owen BB. (1943). *The Physical Chemistry of Electrolytic Solutions*. Reinhold Publishing Corporation: New York.

13.   Head IM, Jones DM, Larter SR. (2003). Biological activity in the deep subsurface and the origin of heavy oil. *Nature* 426: 344–352.

14.   Helgeson HC, Owens CE, Knox AM, Richard L. (1998). Calculation of the standard molal thermodynamic properties of crystalline, liquid, and gas organic molecules at high temperatures and pressures. *Geochim Cosmochim Acta* 62: 985–1081.

15.   Hill RJ, Jarvie DM, Zumberge J, Henry M, Pollastro RM. (2007). Oil and gas geochemistry and petroleum systems of the Fort Worth Basin. *AAPG Bull* 91: 445–473

16.   Jones DM, Head IM, Gray ND, Adams JJ, Rowan AK, Aitken CM*et al.* (2008). Crude oil biodegradation via methanogenesis in subsurface petroleum reservoirs. *Nature*, doi:10.1038/nature06484

17. Larter S, Hockey A, Aplin A, Telnaes N, Wilhelms A, Horstad I et al. (1999). When biodegradation preserves petroleum! Petroleum geochemistry of N. Sea Oil Rimmed Gas Accumulations (ORGAs). *Proceedings AAPG Hedberg Research Conference on 'Natural Gas Formation and Occurrence'*. American Association of Petroleum Geochemists: Tulsa, Oklahoma, pp 3–5.

18. Larter S, Huang H, Adams J, Bennett B, Jokanolaa F, Oldenburgh T et al. (2006). The controls on the composition of biodegraded oils in the deep subsurface: (part II). Geological controls on subsurface biodegradation fluxes and constraints on reservoir fluid property prediction. *AAPG Bull* 90: 921–938

19. Larter S, Wilhelms A, Head I, Koopmans M, Aplin A, Di Primio R et al. (2003). The controls on the composition of biodegraded oils in the deep subsurface—part 1: biodegradation rates in petroleum reservoirs. *Org Geochem* 34: 601–613.

20. Magot M, Ollivier B, Patel BKC. (2000). Microbiology of petroleum reservoirs. *Antonie van Leeuwenhoek* 77: 103–116

21. Manning DAC, Hutcheon IE. (2004). Distribution and mineralogical controls on ammonium in deep groundwaters. *Appl Geochem* 19: 1495–1503

22. Masterson WD, Dzou LIP, Holba AG, Fincannon AL, Ellis L. (2001). Evidence for biodegradation and evaporative fractionation in West Sak, Kuparuk and Prudhoe Bay field areas, North Slope, Alaska. *Org Geochem* 32: 411–441.

23. Nazina TM, Ivanova AE, Borzenkov IA, Belyaev SS, Ivanov MV. (1995a). Occurrence and geochemical activity of microorganisms in high-temperature, water-flooded oil fields of Kazakhstan and Western Siberia. *Geomicrobiol J* 13: 181–192.

24. Nazina TN, Ivanova AE, Golubeva OV, Ibatullin RR, Belyaev SS, Ivanov MV. (1995b). Occurrence of sulfate- and iron-reducing bacteria in stratal waters of the Romashkinskoe oil field. *Mikrobiologiya (Eng Trans)* 64: 203–208.

25. Nazina TN, Shestakova NM, Grigor'yan AA, Mikhailova EM, Tourova TP, Poltaraus AB et al. (2006). Phylogenetic diversity and activity of anaerobic microorganisms of high-temperature horizons of the Dagang oil field (P R China). *Microbiology* 75: 55–65.

26. Orphan VJ, Goffredi SK, DeLong EF, Boles JR. (2003). Geochemical influence on diversity and microbial processes in high temperature oil reservoirs. *Geomicrobiol J* 20: 295–311

27. Orphan VJ, Taylor LT, Hafenbradl D, Delong EF. (2000). Culture-dependent and culture-independent characterization of microbial assemblages associated with high-temperature petroleum reservoirs. *Appl Environ Microbiol* 66: 700–711

28. Pallasser RJ. (2000). Recognising biodegradation in gas/oil accumulations through the $\delta^{13}C$ compositions of gas components. *Org Geochem* 31: 1363–1373

29. Parkes J. (1999). Cracking anaerobic bacteria. *Nature* 401: 217–218

30. Parkes RJ, Webster G, Cragg BA, Weightman AJ, Newberry CJ, Ferdelman TG *et al.* (2005). Deep sub-seafloor prokaryotes stimulated at interfaces over geological time. *Nature* 436: 390–394

31. Roadifer RE. (1987). Size distributions of the World's largest known oil and tar accumulations. In: Meyer RF (ed).*Exploration for Heavy Crude Oil and Natural Bitumen*: AAPG Studies in Geology 25. American Association of Petroleum Geologists: Tulsa, pp 3–23.

32. Rozanova EP, Borzenkov IA, Tarasov AL, Suntsova LA, Dong CL, Belyaev SS *et al.* (2001). Microbiological processes in a high-temperature oil field. *Microbiology* 70: 102–110

33. Rozanova EP, Savvichev AS, Karavaiko SG, Miller YM. (1995). Microbial procsses in the Savuiskoe oil field in the Ob' region. *Mikrobiologiya (Eng Trans)* 64: 85–90.

34. Scott AR, Kaiser WR, Ayers WBJ. (1994). Thermogenic and secondary biogenic gases, San-Juan Basin, Colorado and New-Mexico—implications for coalbed gas producibility. *AAPG Bull* 78: 1186–1209

35. Struchtemeyer CG, Elshahed MS, Duncan KE, McInerney MJ. (2005). Evidence for aceticlastic methanogenesis in the presence of sulfate in a gas condensate-contaminated aquifer. *Appl Environ Microbiol* 71: 5348–5353

36. Sweeney RE, Taylor P. (1999). Biogenic methane derived from biodegradation of petroleum under environmental conditions

and in oil & gas reservoirs. In: Schoell M, Claypool GE (eds). *Proceedings AAPG Hedberg Research Conference on Natural Gas Formation and Occurrence*. American Association of Petroleum Geochemists: Tulsa, Oklahoma, pp 6–10.

37. Thauer RK, Jungermann K, Decker K. (1977). Energy conservation in chemotrophic anaerobic bacteria. *Bacteriol Rev* 41: 100–180

38. Townsend GT, Prince RC, Suflita JM. (2003). Anaerobic oxidation of crude oil hydrocarbons by the resident microorganisms of a contaminated anoxic aquifer. *Environ Sci Technol* 37: 5213–5218

39. Warren E, Bekins BA, Godsy EM, Smith VK. (2004). Inhibition of acetoclastic methanogenesis in crude oil- and creosote-contaminated groundwater. *Bioremediation J* 8: 1–11.

40. Weiner J, Lovley DR. (1998). Rapid benzene degradation in methanogenic sediments from a petroleum-contaminated aquifer. *Appl Environ Microbiol* 64: 1937–1939.

41. Widdel F, Rabus R. (2001). Anaerobic biodegradation of saturated and aromatic hydrocarbons. *Curr Opin Biotechnol* 12: 259–276

42. Wilhelms A, Larter SR, Head I, Farrimond P, Di-Primio R, Zwach C. (2001). Biodegradation of oil in uplifted basins prevented by deep-burial sterilization. *Nature* 411: 1034–1037

43. Zengler K, Richnow HH, Rosselló-Mora R, Michaelis W, Widdel F. (1999). Methane formation from long-chain alkanes by anaerobic microorganisms. *Nature* 401: 266–269.

# Chapter 3

# Life in the Slow Lane; Biogeochemistry of Biodegraded Petroleum Containing Reservoirs and Implications for Energy Recovery and Carbon Management

Ian M. Head[1], Neil D. Gray[1], and Stephen R. Larter[1,2]

[1]School of Civil Engineering and Geosciences, Newcastle University, Newcastle upon Tyne, UK

[2]Petroleum Reservoir Group, Department of Geoscience, University of Calgary, Calgary, AB, Canada

# ABSTRACT

Our understanding of the processes underlying the formation of heavy oil has been transformed in the last decade. The process was once thought to be driven by oxygen delivered to deep petroleum reservoirs by meteoric water. This paradigm has been replaced by a view that the process is anaerobic and frequently associated with methanogenic hydrocarbon degradation. The thermal history of a reservoir exerts a fundamental control on the occurrence of biodegraded petroleum, and microbial activity is focused at the base of the oil column in the oil water transition zone, that represents a hotspot in the petroleum reservoir biome. Here we present a synthesis of new and existing microbiological, geochemical, and biogeochemical data that expands our view of the processes that regulate deep life in petroleum reservoir ecosystems and highlights interactions of a range of biotic and abiotic factors that determine whether petroleum is likely to be biodegraded *in situ*, with important consequences for oil exploration and production. Specifically we propose that the salinity of reservoir formation waters exerts a key control on the occurrence of biodegraded heavy oil reservoirs and introduce the concept of palaeo pickling. We also evaluate the interaction between temperature and salinity to explain the occurrence of non-degraded oil in reservoirs where the temperature has not reached the 80–90°C required for palaeo pasteurization. In addition we evaluate several hypotheses that might explain the occurrence of organisms conventionally considered to be aerobic, in nominally anoxic petroleum reservoir habitats. Finally we discuss the role of microbial processes for energy recovery as we make the transition from fossil fuel reliance, and how these fit within the broader socioeconomic landscape of energy futures.

# AN INTRODUCTION TO HEAVY OIL SYSTEMS

Estimated global bitumen and heavy-oil resources are around 5.6 trillion barrels (bbl), mostly occurring in the western hemisphere. These resources, dominate the world petroleum inventory, and are increasingly being developed as light oil reserves deplete (Hein et al.,

2013). This development has driven applied research into more efficient production, and fundamental research into the biogeochemical origin of these enigmatic large-scale oil accumulations. Much of the enabling technical developments have occurred in the largest bitumen and heavy-oil fields of the Canadian oil sands, the Orinoco Heavy Oil belt of Venezuela, the heavy oil on the North Slope of Alaska and the heavy-oil fields of California. Both applied and fundamental studies, are informed by understanding the microbial processes that have acted on, what began as, more conventional oil and gas. In some cases over geological and, in other cases, more recent timescales.

Heavy oil and bitumen have lower hydrogen and higher carbon contents and hence lower energy content than "conventional," oil and recovery often involves higher energy investment and associated carbon dioxide emissions. Increased oil production from these deposits thus, rightly, raises concerns from society and is now a major political and environmental issue. Perhaps nothing better illustrates the enigmatic and controversial nature of the increasing development of biodegraded oil resources, than the development of Western Canada's oil sands, where output from bitumen reserves, the world's third largest proven crude oil deposit, is expected to climb from around 1.8 million barrels/day in 2012 to 2.3 million barrels/day by 2015, rising to 5.2 million barrels/day by 2030.

# API GRAVITY THE STANDARD MEASURE OF HEAVY OIL

Heavy oil industry terminology is inconsistent and confusing. Based on the oil density-based industry standard yardstick of oil quality— API gravity, many of the "extra heavy oils" of Venezuela would be considered "oil sands" in Canada or "tar sands" in the United States.

API gravity [in degrees] = [141.5/specific gravity at 60F] −131.5)

Heavy oil is defined as oil with 10–20 degrees API and a viscosity of more than 100 centipoise (cP). Bitumen includes extra heavy oil as well as oil in oil sands, with less than 10 degrees API and viscosity of more than 10,000 cP. The main distinction is that the high viscosity of "bitumen," prevents it from flowing to a wellbore under *in-situ* reservoir conditions, whereas heavy oils will flow. Heavy oil and bitumen can be regarded as part of a continuum of heavily to severely biodegraded oil (Hein et al., 2013).

# GEOLOGICAL SETTINGS AND PROPERTIES OF HEAVY OIL AND HEAVY OIL RESERVOIRS

Although, heavy oil can be found in all basin styles and sizes, from narrow rift basins to the largest sedimentary basins on Earth, most of the largest heavy oil and oil sands deposits are found in large foreland basins adjacent to orogenic belts, with large source rock kitchens charging large, shallow, cool, reservoirs at the basin flank susceptible to severe *in-situ* biodegradation (Creaney et al., 1994; Adams et al., 2013a). The world's largest oil-sand deposit, located in western Canada, is reservoired in Lower Cretaceous sandstone deposits in a basin adjacent to the Canadian Rocky mountains (a foreland basin) (Head et al., 2003; Adams et al., 2004, 2013a; Larter et al., 2006; Larter and Head, 2014). Petroleum was derived principally from marine shale source rocks with the petroleum migrating eastward several hundred kilometers to accumulate and biodegrade on northeastern margins of the basin. The main phase of accumulation was around 84–55 Ma ago (Adams et al., 2013a; Tozer et al., 2014). The petroleum accumulated in tidal-controlled river and estuarine sediments. Oil was similarly accumulated in foreland basin settings in the Oficina Formation in Venezuela, another major heavy-oil resource.

The primary control on oil composition and viscosity in heavy oil and bitumen containing reservoirs, is in-reservoir, anaerobic biodegradation (Head et al., 2003; Larter et al., 2008). The mechanisms, control, and consequences of this are discussed in detail later in this manuscript. Alberta Lower Cretaceous reservoirs range from 38°API (light oil) in the barely biodegraded oil pools west of the Peace River oil sands, to 6° API (severely biodegraded bitumen) in eastern Athabasca oil sands with even lower values in the most degraded bitumens, found in karsted Grosmont carbonate reservoirs underlying the oil sands. Oil sulfur contents range from 1 to >10 wt%, with the western Peace River oil sands having the highest values. Variability correlates roughly to levels of oil biodegradation which broadly increase from west to east and from south to north (Adams et al., 2006).

The impact of the deep biosphere on petroleum composition is significant since biodegradation affects oil composition which in turn affects fluid flow properties (viscosity), and oil pricing (API gravity), both factors with enormous economic and environmental impact. Oil compositional gradients and resulting vertical and lateral oil viscosity variations (discussed below) are common on both reservoir thickness (tens of meters) and field scales (kilometers) and are a defining characteristic of heavy oilfields. Such gradients in a minority of heavy oilfields can certainly be produced by restricted vertical mixing and by density stratification of an evolving oil charge, as originally suggested by Khavari-Khorasani et al. (1998) and more recently by Stainford (2004). Importantly it is in-reservoir oil biodegradation that substantially produces the systematic compositional gradients seen in heavy oilfields (Larter et al., 2003). Furthermore, continuous vertical compositional gradients in the oil columns document episodic degradation for many millions of years, suggesting that the timescales of oilfield degradation and petroleum charging are similar (Larter et al., 2003). Gradients in chemical composition of oil commonly seen in heavy oilfields, include differences in the relative and absolute concentrations of compounds such as n-alkanes, and isoprenoid alkanes which are relatively easily degraded components and typically decrease in concentration toward oil-water contacts. As discussed below, oil-water contacts are the primary site of oil biodegradation.

By contrast cyclic biomarker alkanes such as hopanes are more resistant to biodegradation and often increase in concentration toward the oil-water contact at low and intermediate levels of degradation. There are also, systematic molecular changes in the relative abundance of multiple components with very similar densities through the oil columns. This includes for example, decreasing isoprenoid alkane abundances relative to hopanes, or regular sterane abundances relative to diasteranes (Peters and Moldowan, 1993; Peters et al., 2005) which indicate the biodegradation process is driven by molecular selectivity rather than by gravitational processes related to compound density. Other examples include selective removal of individual aromatic hydrocarbon isomers of similar density toward the bottom of oil columns (e.g., alkylphenanthrenes, napththalenes, or dibenzothiophenes), which unambiguously indicate that biodegradation, not gravitational effects control oil composition variations in heavy oilfields.

While some degree of water washing may also have taken place penecontemporaneously (almost at the same time), with biodegradation, selective removal of specific alkylaromatic hydrocarbon or non-hydrocarbon isomers, with similar water solubilities (Taylor et al., 2001), suggests that biodegradation is dominant as the alteration process, and water washing is of little significance.

In this hypothesis and theory article we propose new ideas and explore emerging paradigms about the controlling factors that dictate in-reservoir petroleum biodegradation and the microbial ecology of heavy and other petroleum systems, and offer a perspective on the implications for energy recovery and carbon management in the future.

# CONTROLS ON IN-RESERVOIR PETROLEUM BIODEGRADATION

Field observations typically record the lowest oil quality and the strongest biological and molecular evidence for hydrocarbon degradation at oil-water transition zones (OWTZ), suggesting that most petroleum degradation occurs at this interface (Head et al., 2003; Larter et al., 2003); where the biosphere meets the geosphere. This conclusion is sensible because at the oil-water contact, organisms find the water necessary for life along with electron donors and acceptors and carbon sources (oil) necessary to conserve energy and generate biomass. Essential nutrients, such as nitrogen and phosphorus are primarily derived from mineral buffering in the water leg (cf. Rogers et al., 1998; Head et al., 2003; Bennett et al., 2013).

Multiple and complex oil-water contacts may exist throughout a charging oil reservoir. However, the large vertical compositional gradients in oil columns of degraded oil reservoirs with the most degraded oils at the oil-water contact (Moldowan and McCaffrey, 1995; Horstad and Larter, 1997; Huang et al., 2003; Larter et al., 2003; Bennett et al., 2013) suggest the bulk of the biodegradation in most reservoirs is ultimately driven from the basal oil-water contact of the filling reservoir. Although petroleum biodegradation undoubtedly occurs in shallow carrier systems too, relative residence time in-reservoir, compared to in-carrier, is large, suggesting reservoir processes dominate alteration of oil.

The past decade has witnessed fundamental changes in our concept of in-reservoir petroleum biodegradation with a shift from a model that emphasized the central importance of aerobic oil degradation, with oxygen delivered to petroleum reservoirs in meteoric water (Palmer, 1993) to a model where in-reservoir crude oil biodegradation is driven by anaerobic processes. In reservoirs with low concentrations of sulfate, methanogenic degradation is a primary mechanism of petroleum biodegradation (Head et al., 2003; Larter et al., 2005; Jones et al., 2008). This conclusion was based on field measurements of metabolites characteristic of anaerobic hydrocarbon degradation (Aitken et al., 2004), comparative analysis of field-degraded oils with oils degraded under methanogenic or sulfate-reducing conditions in laboratory incubations, analysis of gas isotopes from the field and Rayleigh fractionation modeling (Jones et al., 2008). This conceptual shift has important implications for understanding the factors that control biodegradation *in situ*, and therefore, for establishing the likelihood that a prospect will be biodegraded pre-drill. Moreover, the mechanisms whereby petroleum hydrocarbons are degraded may dictate which diagnostic metabolites might act as signatures of different processes occurring in relation to prevailing reservoir conditions (e.g., low sulfate clastic systems vs. high sulfate carbonate systems). The following sections detail the factors affecting anaerobic and specifically methanogenic crude oil biodegradation.

# OIL PROPERTIES

Here we evaluate the effects of different oil components on the methanogenic degradation of crude oil using both observations from field biodegraded oils and experimental approaches.

## Do Water Soluble Oil Components Retard Methanogenic Degradation of Crude Oil?

It has been suggested that reservoir geometry has an important effect on the extent of biodegradation of crude oil in petroleum reservoirs

(Larter et al., 2005, 2006). This has been interpreted principally in relation to the availability of water and supply of inorganic nutrients from the water leg, controlled by the area of the oil water contact in relation to the size of the reservoir and the size of the underlying aquifer. However, this could also be interpreted in terms of water washing with removal of inhibitory, water soluble components of the oil such as phenols. It has been suggested that petroleum reservoirs are biostatic (Sunde and Torsvik, 2005) i.e., biological activity is inhibited but the organisms present are not killed. Biostatic conditions in petroleum reservoirs have been explained in terms of toxic water soluble hydrocarbons or other components of the oil such as metals. It might also relate to toxic low molecular weight, volatile components of crude oil as inhibitors (see Section Do Volatile Hydrocarbons Influence Methanogenic Degradation of Crude Oil?). The suggestion that oil reservoirs are biostatic is based on the common observation that while oilfield waters often contain sulfate-reducing bacteria, electron donors, oxidants and nutrients, spontaneous $H_2S$ production is not observed until the oilfield water is degassed or diluted, when, $H_2S$ production subsequently takes place (Sunde and Torsvik, 2005). The biostatic nature of reservoir waters is paradoxical in the context of demonstrable in-reservoir oil biodegradation and therefore would imply a further control on oil biodegradation *in situ* which may relate to the activity of any associated aquifers. For instance reservoirs associated with an active aquifer where large volumes of water pass the oil water contact will be biodegraded, as toxic, water soluble oil components would be washed from the oil. This hypothesis would also be consistent with the lack of oil biodegradation in reservoirs with an under seal with oil completely filling the reservoir compartment down to the under seal. At field scale, it is not clear how the effects of removing or diluting toxic water soluble compounds could be distinguished from enhanced delivery of nutrients by a more active aquifer. Interestingly, new results presented here show that water washing of oil has little effect on rates of methanogenesis in laboratory incubations (Figure 1). From these results it seems unlikely that readily water-extractable toxic polar compounds in oil inhibit methanogenic crude oil biodegradation or explain the prevalence of methanogenic $CO_2$ reduction sometimes seen during crude oil biodegradation and described as MADCOR (Methanogenic Alkane Degradation Dominated by $CO_2$ Reduction; Jones et al., 2008). A possible explanation of MADCOR is preferential inhibition of acetoclastic methanogens (Warren et al., 2003, 2004; Westerholm et al., 2011). Interestingly Warren et al. (2003, 2004) noted that aqueous extracts of creosote were inhibitory to acetoclastic methanogenesis. The amounts of oil used in the laboratory experiments reported here were small (Figure 1). In general in many petroleum systems, oil contacts relatively small volumes of water compared with the volume of the reservoir, and this also suggests that water soluble inhibitors are in reality not a primary factor in driving MADCOR.

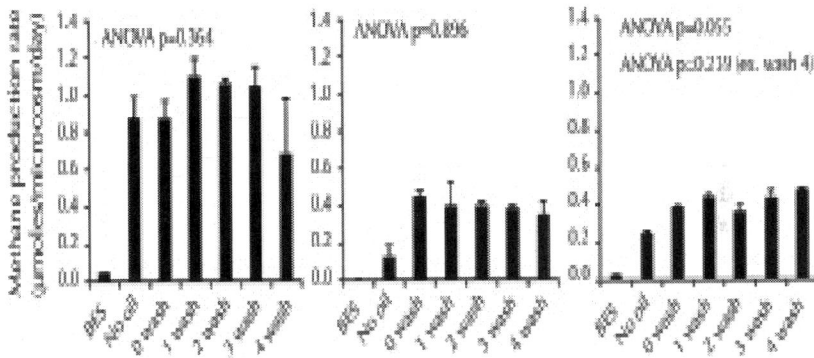

**Figure 1:** The effect of oil and water-washed oil on rates of methanogenesis from $H_2/CO_2$ (left), acetate (middle), and methanol (right). Laboratory microcosms amended with oil which had been washed with brine to different degrees was evaluated. North Sea oil (50 g) was washed with 100 ml brine (1% w/v NaCl in deionized water) in a separating funnel from 1 to 4 times and the unwashed and washed oil (300 mg) was added to serum bottles containing a slurry of river sediment (River Tyne, UK) amended with each of the methanogenic substrates. Control incubations containing bromoethanesulfonic acid (BES), an inhibitor of methanogenesis were also prepared. Rates of methane production were measured by gas chromatography of headspace samples (E. Bowen, previously unpublished data).

# Do Volatile Hydrocarbons Influence Methanogenic Degradation of Crude Oil?

Several studies indicate that low molecular weight hydrocarbons inhibit the aerobic biodegradation of crude oil (Westlake et al., 1974; Atlas, 1975; Walker et al., 1976; Wang et al., 1998). This may explain why aerobic biodegradation of spilled oil is not observed until volatilization has removed low molecular weight alkanes (Atlas and Bartha, 1972b; Whittle et al., 1982). Moreover there is evidence that some, but not all anaerobic bacteria are inhibited by short chain alkanes (<C9; Rodriguez-Martinez et al., 2008). This may also explain the observation of biostatic conditions that prevent sulfate reduction in petroleum reservoirs.

Interestingly a comparison of methanogenic oil-degrading microcosms provided with a North Sea oil containing volatile aromatic and aliphatic hydrocarbons with the same oil from which the volatile components (alkanes < C11 and monoaromatic hydrocarbons) were removed (by incubation of the oil at 25°C for 48 h) demonstrated that methanogenic oil degradation occurred after a much shorter time period, and at a much faster rate in microcosms lacking the volatile hydrocarbons (Sherry et al., 2014). Nevertheless oil degradation was not completely inhibited in the presence of the volatile hydrocarbons and thus may have an important role to play in modulating the rate of in-reservoir crude oil biodegradation.

# ENVIRONMENTAL CONSTRAINTS ON ANAEROBIC CRUDE OIL BIODEGRADATION

A number of environmental factors are known to affect microbial activity, these include temperature, pH, salinity, water activity, radiation and availability of resources such as carbon and energy sources, electron acceptors and inorganic nutrients. For in-reservoir crude oil biodegradation the most important of these are temperature, salinity, and inorganic nutrient availability.

## Temperature

Connan (1984) first suggested that in-reservoir oil biodegradation ceased above a reservoir temperature of around 80°C. However, non-degraded reservoirs are also found at lower temperatures and an explanation for their occurrence came with the development of the palaeopasteurization hypothesis, developed at the turn of the twentieth century (Wilhelms et al., 2001). The palaeopasteurization hypothesis, formulated based on field data, indicated that the upper thermal limit for hydrocarbon-degrading microbial life in petroleum reservoirs was 80–90°C and that once a reservoir had been heated to temperatures within this range, it was not re-colonized by hydrocarbon degrading microorganisms, even if the reservoir was subsequently uplifted to shallower cooler depths (Wilhelms et al., 2001). Subsequently a systematic relationship between biodegradation, reservoir depth, and burial temperature was observed in the Western Canada Sedimentary Basin (Adams et al., 2006). Thus, temperature is considered a primary control on the occurrence of biodegraded petroleum reservoirs.

From the literature it appears that thermophilic hydrocarbon-degrading microorganisms are rare. One moderately thermophilic alkane-degrading sulfate reducing bacterium (*Desulfothermus naphthae* strain TD3), with optimum activity at 60°C has been isolated from Guaymas Basin sediments (Rueter et al., 1994), which degraded low molecular weight alkanes. It is only recently that hydrocarbon degradation at relatively high temperatures has been reported in laboratory incubations inoculated from petroleum reservoirs under both sulfate-reducing and methanogenic conditions (Gieg et al., 2010; Mbadinga et al., 2012; Zhou et al., 2012). These findings are supported by studies of pure and enrichment cultures that demonstrate the hydrocarbon-degrading ability of a range of thermophilic and hyperthermophilic anaerobic bacteria and archaea. Chen and Taylor (1997) reported the degradation of some BTEX hydrocarbons at 50°C by a sulfate-reducing enrichment culture. Additionally, propane degradation at 60°C occurred in a sulfate-reducing enrichment dominated by a spore-former related to *Desulfotomaculum* (Kniemeyer et al., 2007) and short chain alkane degradation linked to sulfate reduction occurs in hydrothermal sediments at temperatures up to 75°C (Adams et al., 2013b).

Hyperthermophilic archaea are also now known to degrade hydrocarbons. Isolates of the sulfate-reducing archaeon *Archaeoglobus fulgidus* have been obtained from hydrothermal systems and petroleum reservoirs with some indications that isolates from different habitats are very similar at the genomic level (Stetter et al., 1993). Early evidence showing that *Archaeoglobus* could grow on crude oil indicated that it was most likely fatty acids in the oil that served as growth substrates

(Stetter et al., 1993). However, it was subsequently demonstrated that the type strain of *A. fulgidus*(strain VC 16) could grow on alkenes (Khelifi et al., 2010) and long chain alkanes (Khelifi et al., 2014) at 70°C. Another member of the *Archaeoglobales, Ferroglobus placidus*, degrades benzene with ferric iron as an electron acceptor at 85°C. This is the current upper limit for anaerobic hydrocarbon degradation by a pure microbial culture (Holmes et al., 2011). There is also some evidence to suggest that *Thermococcus sibiricus* may have the capacity to degrade alkanes (Mardanov et al., 2009).

While these studies bring hydrocarbon degradation into the realm of the hyperthermophiles, data remain consistent with the upper temperature limit for hydrocarbon-degrading life in the deep petroleum biosphere inferred from the palaeopasteurization hypothesis (Wilhelms et al., 2001). There are few reports of deep subsurface hyperthermophiles that have been isolated at temperatures greater than 90°C when oilfield waters have been used as an inoculum (Grassia et al., 1996) and, interestingly, methanogenesis and sulfate reduction could be measured in produced waters from Californian petroleum reservoirs only at temperatures between 70 and 83°C and not at higher temperatures, even when the temperature of the reservoir from which the samples came, was up to 120°C (Orphan et al., 2003). It is also of interest that the organisms implicated in hydrocarbon degradation around the field-derived palaeopasteurization threshold, use electron acceptors offering relatively high energy yields and there have to date been no reports of methanogenic oil degradation at such high temperatures. It is also of note in this regard, that *A. fulgidus* can degrade alkenes with either sulfate or thiosulfate as an electron acceptor (Khelifi et al., 2010), while degradation of alkanes was only reported with thiosulfate, and not the less energetically favorable sulfate, as an electron acceptor (Khelifi et al., 2014).

Observations from laboratory and field measurements are also informative. Re-examination of data from our own laboratory shows that methanogensis from oil degradation (but not from indigenous organic matter) is inhibited by pasteurization (Figure 2). This suggests hydrocarbon fermenting bacteria which provide methanogens with hydrogen, carbon dioxide and acetate are more heat-sensitive than the methanogens themselves since pasteurized sediments containing crude oil continue to produce methane but only at levels equivalent to control incubations to which no oil has been added (Figure 2).

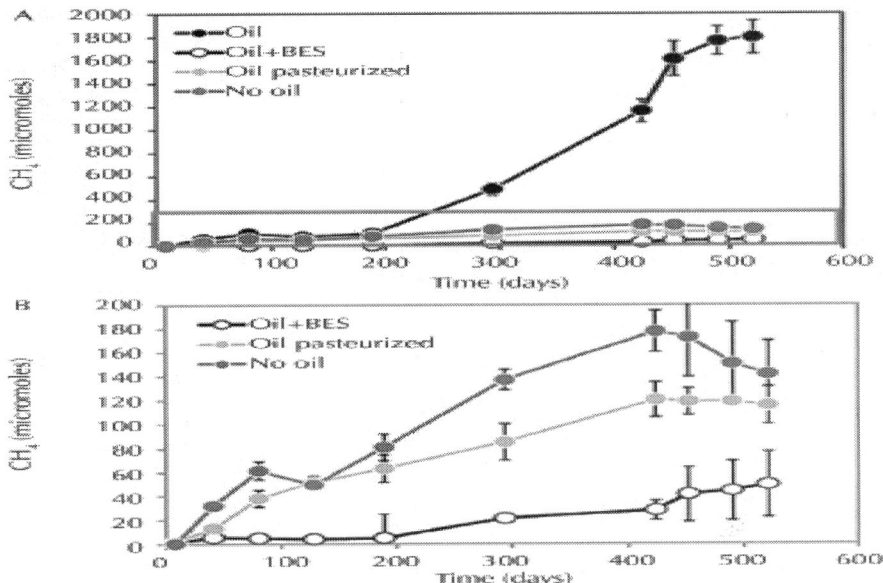

**Figure 2:** Effect of pasteurization in methanogenesis and methanogenic oil biodegradation in laboratory microcosms. Panel (B) is a blow up of the red box in (A). This demonstrates that following pasteurization (90°C for 2 h) methanogenesis attributable to oil biodegradation ("Oil" in A) is reduced. The data from pasteurized control microcosms amended with oil suggest that methanogens remain active, even after pasteurization has removed hydrocarbon degrading activity. This is evident from the fact that pasteurization does not completely inhibit methanogenesis but brings methane production levels in line with that seen in background control incubations (No oil) to which no oil was added suggesting that the capacity for conversion of alkanes to methane has been removed by pasteurization, but not the ability to generate methane from indigenous organic carbon in the sediments (A. Rowan, previously unpublished data).

A lower thermal maximum for microbial life in petroleum systems is also evident from measurements of methanogenesis from complex organic matter (yeast extract) in water samples from petroleum reservoirs in the Western Canada Sedimentary Basin that span a temperature gradient from 67 to 113°C (Figure 3). Methanogenesis could be measured in almost all samples from reservoirs with *in situ* temperatures below 90°C but only in one reservoir above this temperature (previously unpublished data, Figure 3). Furthermore, it was possible to recover intact bacterial polar lipids (IPL) only from lower temperature reservoirs (Oldenburg et al., 2009). A number of lines of evidence therefore suggest that in petroleum reservoirs the maximum temperature for biological activity is below that observed for cultivated organisms isolated from high temperature hydrothermal systems (121–122°C; Kashefi and Lovley, 2003; Takai et al., 2008). Nevertheless, intact prokaryotic cells have been identified in deep sediments where temperature exceeds 80–90°C. Mud volcano breccia inferred to emanate from depths where the temperature is estimated to be up to 160°C (Parkes et al., 2000), sediments from the Newfoundland Margin [1626 mbsf, 111 My and 60–100°C (Roussel et al., 2008)] and deep coal seams in New Zealand (estimated maximum temperature 80–90°C; Fry et al., 2009) have all been shown to harbor large numbers of prokaryotic cells, though their activity, and specifically their hydrocarbon degradation potential, at these temperatures is unknown.

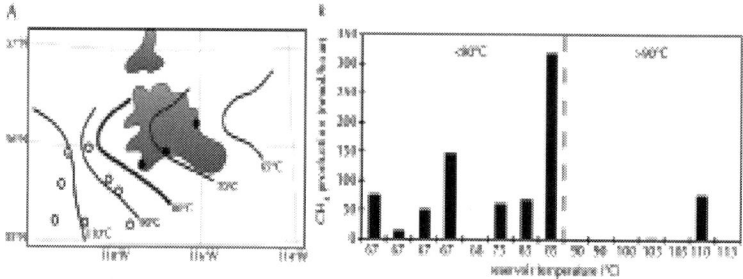

**Figure 3:** Oil production wells (circles) near the Peace River tar sands (gray shaded areas) in Alberta (A). Isotherms indicate *in situ* temperatures of crude oil deposits. Consistent with palaeopasteurization, we recently recovered microbial intact polar lipids in produced water samples from wells east of the 80°C isotherm (●) but not from more western reservoirs (○) that are above 80°C (Oldenburg et al., 2009). Panel(B) shows that microbial activity (methanogenesis from yeast extract) was readily stimulated in samples from eastern fields lower than 80–90°C *in situ* (dashed line) but not from deeper, hotter fields to the west (N. Gray, unpublished data).

One reason for a modulated maximum thermal limit for life in some deep subsurface sediments may be that the rates of metabolism that can be sustained in a metabolically constrained system do not allow for the regeneration of labile molecules involved in conservation of energy (e.g., ATP and NADH) sufficiently quickly to support the maintenance energy requirements of cells (Wilhelms et al., 2001). It has been shown that cells with low metabolic activity have reduced tolerance to environmental extremes (Lloyd et al., 2005). This may be exacerbated by additional energy requirements to maintain cell integrity in a chemically demanding environment such as crude oil. There are many other extremes found in deep subsurface sediments that may also conspire to limit the tolerances of microbial life and it is known that interactions between different environmental factors can act to increase or decrease an organism's sensitivity to other environmental stressors (Edgcomb et al., 2004).

# Salinity

It is well known that salinity affects microbial activity and that petroleum reservoir formation waters can vary from freshwater, to many times seawater salinity. Reservoir water salinity will therefore affect the occurrence of biodegraded oils. Furthermore, it is likely that there will be an interaction between salinity and other environmental factors such as temperature, and for example, the temperature required for palaeopasteurisation of a reservoir may be lower if the salinity is elevated. Although there is no systematic analysis of salinity-biodegradation relationships in oil fields, there is anecdotal evidence that some low temperature, non-uplifted, non-biodegraded reservoirs are associated with high salinity aquifers. We therefore propose that "Palaeopickling" (cf. palaeopasteurisation) may be a mechanism whereby biodegradation of crude oil is prevented, providing a further factor that should be considered for pre-drill prediction of biodegradation.

## Salinity Tolerances and Mechanisms of Osmoadaptation in Microorganisms

Microorganisms are known that can grow over the full salinity range from almost pure water to saturated brines and particular organisms are adapted to particular ranges of salinity (DasSarma and Arora, 2002). To counteract the effect of osmotic efflux of water in high solute environments microorganisms have evolved two main mechanisms;

- accumulation of ions (typically K+ and Cl− ions) to high intracellular concentrations (the "salt in" strategy) and
- intracellular synthesis of organic compounds known as compatible solutes (so called because they are compatible with the activity of enzymes).

Both the "salt in" and solute synthesis strategies require considerable energy expenditure with the synthesis of compatible solutes being energetically more demanding than accumulation of ions. This energy cost must be deducted from the energy yield from primary metabolism. This typically means that the upper salinity limit for growth and activity is lower for organisms that live by harnessing reactions that have low energy yields (Oren, 2010). Although, low energy yield *per se* does not preclude activity at higher salinity, the rate at which energy can be supplied does. Given a finite metabolic activity, the point at which energy can no longer be supplied at a sufficient rate to service cell maintenance, dictates when a microbial process will cease.

Synthesis of compatible solutes takes from 20 to >100 moles of adenosine triphosphate (ATP) per mole of compatible solute depending on the specific molecule (Oren, 1999). In principle the internal concentration of the compatible solute must be twice the external salt concentration (NaCl dissociates into 2 ions). In practice this value is lower due to the osmolarity contribution from normal intracellular solutes including proteins but still represents a considerable metabolic burden on a cell which is linearly proportional to the salinity. We have estimated that the ATP requirement per cell increases by an order of magnitude between 0 and 400 g/L salt concentration (Figure 4). The salt in strategy is less energetically expensive, but still exerts a similar proportional increase in burden over this range (Figure 4).

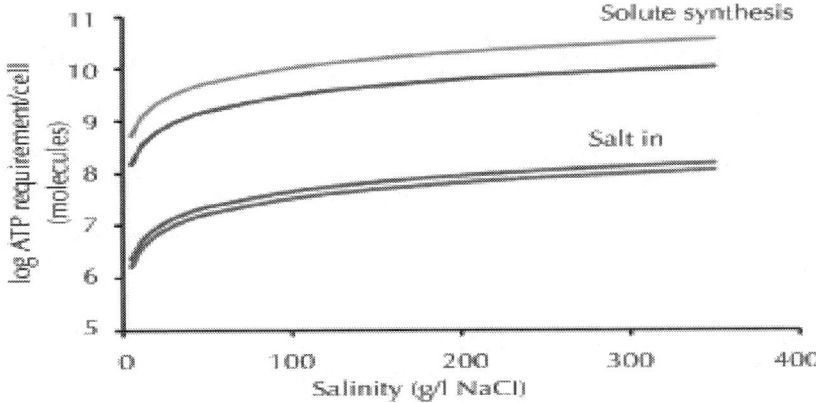

**Figure 4:** Effect of increasing salinity on energy (ATP) requirement/cell for osmoregulation. The upper and lower lines in each pair of lines relates to the upper and lower values quoted by Oren (1999). Calculations were based upon the following assumptions; Intracellular solute concentration is regulated to match the extracellular salinity; for organic solutes 2 molecules are required per salt molecule (1 molecule of salt produces 2 ions); Cells are cocci (spheres) of 0.5 microns in diameter. Note log scale, thus there is a linear increase in energy demand with increasing salinity.

Salinity effects on methanogenesis, the terminal process in methanogenic oil biodegradation depends on the specific pathway utilized, with tolerance increasing with increasing substrate energy yield. Typically, acetoclastic methanogenesis yields least energy and has a relatively low upper salinity limit (Oren, 2010). Methanogenesis from $H_2 + CO_2$ has a higher salinity tolerance but has never been shown above 175 $gl^{-1}$ salt (Zhilina et al., 2013). At higher salt concentrations, biogenic methane comes from disproportionation of methylated amines which yield more energy than other methanogenic substrates. Of course the specifics will be dictated by *in situ* conditions such as acetate or hydrogen concentration which will affect metabolic rates and the overall energy available in the system.

## *Salinity Effects in the Context of Methanogenesis in Petroleum Reservoirs*

It has been noted that the frequency of isolation of microbial cultures from petroleum reservoirs decreases above a salinity of ca. 100 g/l

(Grassia et al., 1996; Röling et al., 2003). Furthermore, methanogenesis from reduction of $CO_2$ with $H_2$, the most prevalent methanogenic process seen in petroleum reservoirs (Head et al., 2010), has never been detected in natural environments at salt concentrations exceeding 100 g/l. This is consistent with the observation of methanogenesis from $CO_2$ with $H_2$ only at salinities up to 90 g/l in a high temperature, saline North Sea oil and gas reservoir, where the *in situ* salinity was 90 g/l (Gray et al., 2009).

However, the cultured $H_2$-oxidizing $CO_2$ reducing methanogen, *Methanocalculus halotolerans*, was, isolated from an oil well and grows at salt concentrations up to 120 g/l NaCl and at temperatures up to 45°C (Olivier et al., 1998). It is likely that the discrepancy between the salt tolerance of methanogenesis by *Methanocalculus halotolerans* and *in situ* methanogenesis rates is the fact that the maximum growth temperature of *Methanocalculus halotolerans* is somewhat lower than the *in situ* reservoir temperature and it is possible that at higher temperatures the salinity tolerance of this methanogen may be lower. This is the first circumstantial evidence that there may be an interaction between salinity and temperature on microorganism activity and the possibility that palaeopickling of low temperature reservoirs with very saline formation waters may prevent biodegradation.

We have conducted a systematic analysis of the effect of salinity on rates of methanogenesis and report the results here. This was done using river sediment microcosms to which different levels of NaCl were added to give a range from 1 to 137 g/l NaCl and these were amended with a range of methanogenic substrates. Two sets of incubations were conducted, one at 30°C and one at 60°C, and rates of methane production were measured. These experiments clearly demonstrated that there was an interaction between temperature and salinity with the salinity tolerance of methanogenesis from all methanogenic substrates being greater at 30°C than at 60°C. At 30°C, the maximum salt concentration permissive for methanogenesis from acetate, $CO_2$ and $H_2$, and methanol was 46, 114, and 137 g/l NaCl respectively (Figure 5), confirming the observations of Oren (2010). However, at 60°C methanogenesis from all substrates ceased above 46 g/l NaCl (Figure 5).

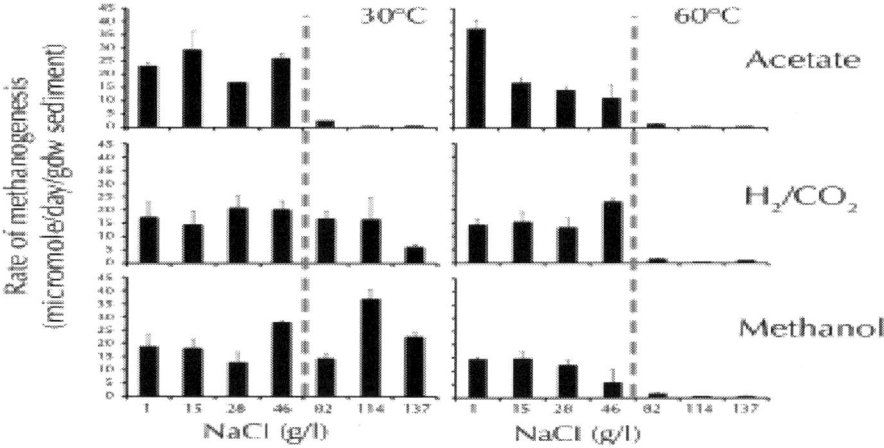

**Figure 5:** Methanogenesis from acetate, hydrogen and $CO_2$ and methanol in anoxic incubations of River Tyne sediments incubated at 30°C (left panel) and 60°C (right panel) over a range of salt concentrations from 1 to 137 g/l. The maximum salt tolerance for methanogenesis is lower at higher temperature indicating an interaction between salinity and temperature (H. Coombs, previously unpublished data). At 60°C methanogenic activity is inhibited above 46 g/l indicated by the dashed vertical red line.

Clearly the terminal processes in crude oil degradation under methanogenic conditions show an interaction between salinity and temperature that may affect the occurrence of in-reservoir oil biodegradation. How might this translate into effects on biodegradation of oil? As indicated above, the energetic cost of osmoadaptation is high and proportional to the salt concentration and thus processes that yield less energy are likely to be compromised by high salinity conditions (Oren, 2010). These are the very processes that are central to methanogenic oil biodegradation. In addition to acetoclastic methanogenesis, syntrophic proton reducers i.e., primary alkane fermenters and syntrophic acetate oxidizers, yield small amounts of energy and have relatively limited salinity tolerance. These processes coupled to methanogenesis are the principal processes in methanogenic degradation of crude oil alkanes (Gieg et al., 2008; Jones et al., 2008). Taken together these data and observations suggest that palaeopickling may be important for preserving oil in low temperature petroleum reservoirs and that the palaeopasteurization temperature required to preserve oil may be lower in high salinity systems, potentially providing a further approach to inform pre-drill prediction of biodegradation. We predict that at higher salinity the palaeopasteurization temperature is likely to be reduced from the maximum field observed value of 80°C. However, the precise relationship between salinity, temperature and in-reservoir oil preservation is currently unknown. A systematic analysis and calibration against field data from petroleum reservoirs is therefore much needed.

# Nutrient Availability

A primary control on the degradation of crude oil in *aerobic* environments is nutrient availability (Atlas and Bartha, 1972a; Head et al., 2006). Degradation relies on the growth of hydrocarbon-degrading organisms and hydrocarbon conversion to biomass for growth requires additional nitrogen and phosphorus. These nutrients make up a significant proportion of biomass and are typically around 15% N and 1% P on a dry weight basis (Redfield, 1934, 1958) though this does vary (e.g., Fagerbakke et al., 1996; Gunderson et al., 2002; Vrede et al., 2002). Thus, to convert hydrocarbons to biomass, supplementary N and P are required. This is one of the reasons that nutrient availability is also considered an important control on in reservoir oil biodegradation. However, anaerobes typically have lower biomass yields than aerobes and growth yields of methanogenic hydrocarbon degrading consortia are in the range of a 2–10% compared to the 50% growth yield expected for aerobic heterotrophs (Gray et al., 2011). Thus, in methanogenic oil degrading systems such as biodegrading petroleum reservoirs proportionately less N and P are required to permit degradation of a given mass of hydrocarbon coupled to growth.

There are few data on inorganic nitrogen concentrations in deep biosphere sediments. Relatively shallow (219 m) and young sediments (4,60,000 y) from the northwestern Pacific Ocean contained ammonium concentrations of 15 mM and based on stable isotope tracer studies with [15]N-labeled ammonium it was concluded that growth and activity were energy- and not inorganic nutrient-limited (Morono et al., 2011). Typically deep subsurface sediments are considered carbon and energy limited as organic carbon quality and quantity decrease with depth and microbial biomass correlates with organic carbon content (D'Hondt et al., 2009). Petroleum reservoirs are different as they contain large quantities of potential electron donor. Deep groundwater from petroleum reservoirs may contain substantial amounts of ammonium [up to 1000 mg/l (55 mM); Manning and Hutcheon, 2004]. Intriguingly, in a study of oil sand reservoirs, conservative estimates of formation water N and P indicated that cell counts, on the order of those observed in formation waters (ca $10^5$–$10^6$ per gram), can be supported without removing additional nutrients from the oil or surrounding rock thus N and P availability does not appear to constrain the degree of oil biodegradation in this system.

While inorganic nitrogen is the likely immediate source of nutrients for the in-reservoir oil biodegradation system, petroleum contains considerable amounts of organic nitrogen. Total nitrogen content of crude oils range between 0.01 and 0.9% by weight (Ball et al., 1951). However,Richter et al. (1952), suggested that most of the organic nitrogen is present in the high molecular weight fraction of crude oils. The low molecular weight fraction of crude oil, in terms of its nitrogen compounds is dominated by alkylcarbazoles, and these are biodegraded at degradation levels above Peters and Moldowan level 4, and might supply nitrogen to microbial communities resident in petroleum reservoirs (Oldenburg et al., 2006). Clearly, alkylcarbazole-derived nitrogen is not used during the early stages of biodegradation when alkane removal is the primary process occurring. Thus, oil-derived nitrogen may not be quantitatively critical at the times when the highest rates of hydrocarbon removal occur.

# THERMODYNAMIC DRIVERS

The accumulation of metabolic intermediates and end products of oil degradation may exert a thermodynamic control on in-reservoir crude oil biodegradation linked to methanogenesis.

## Inhibition by Accumulation of Metabolic Intermediates ($H_2$)

Methanogenic alkane degradation may proceed via several routes that can involve complete oxidation to hydrogen and carbon dioxide, fermentation to acetate and hydrogen and syntrophic acetate oxidation

(Dolfing et al., 2008). Hydrogen is an important intermediate in these processes (Dolfing et al., 2008) and an examination of the effect of hydrogen partial pressure on the theoretical Gibb's free energy yield of each of these processes confirms that for the conversion of alkanes to methanogenic substrates a very low hydrogen partial pressure (less than 4 Pa/ $4 \times 10^{-5}$ atm) is required if these processes are to remain thermodynamically feasible (Figure 6).

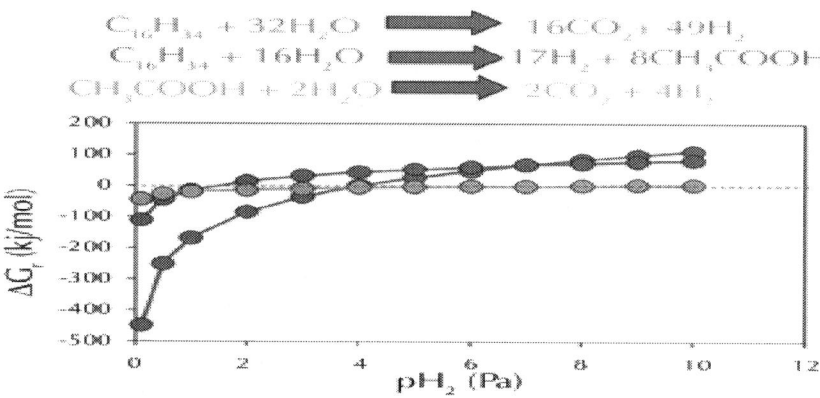

**Figure 6:** The effect of hydrogen partial pressure on the free energy yield from oxidation of hexadecane to hydrogen and $CO_2$ (blue), fermentation of hexadecane to hydrogen and acetate (red) and syntrophic acetate oxidation (green).

These low levels of hydrogen are typically maintained by methanogens that efficiently consume hydrogen produced by fermentation reactions, however if their activity is reduced due to inhibition by some environmental factor such as pH or presence of inhibitory compounds, then partial pressure of hydrogen can rise leading to a cessation of primary fermentation of the alkanes. Clearly then accumulation of metabolic intermediates such as hydrogen potentially has a profound effect on methanogenic crude oil biodegradation.

# Inhibition by Accumulation of Metabolic End-products ($CH_4$ and $CO_2$)

It is relatively well known from studies of methanogenic systems that accumulation of metabolic intermediates such as hydrogen leads to thermodynamic inhibition of biodegradation of many compounds. However, it is usually assumed that metabolic end products have little effect on the thermodynamics of methanogenic systems. Using standard conditions except for methane concentration, we have calculated that methanogenic hexadecane degradation remains thermodynamically feasible even at methane concentration greater than $10^4$ atmospheres (Figure 7). Carbon dioxide concentrations are more permissive and the free energy change remains negative up to $CO_2$ concentrations of $10^{17}$ atmospheres (Figure 7). These concentrations far exceed those likely to be attained in biodegrading petroleum systems. If methane and $CO_2$ concentrations are considered together and a "window of opportunity" defined, where methanogenic alkane degradation is thermodynamically favorable (Dolfing et al., 2008) the window of opportunity is huge.

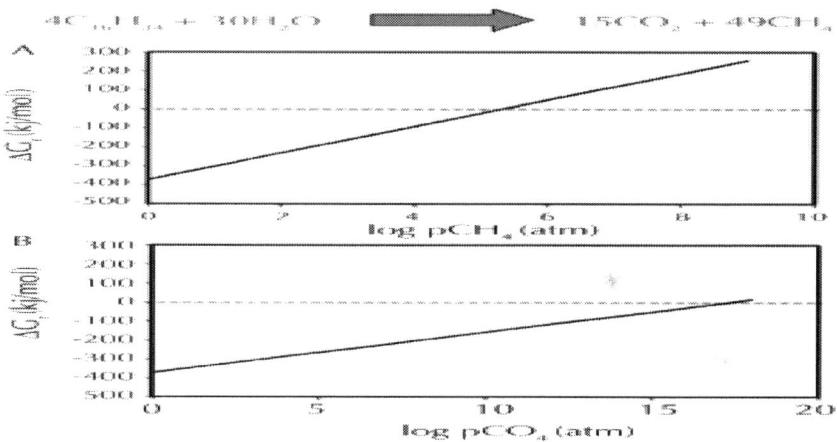

**Figure 7:** Free energy yield from methanogenic hexadecane degradation at a range of methane (A) and carbon dioxide (B) partial pressures.

Methane and $CO_2$ partial pressures likewise, are unlikely to have a major impact on the thermodynamic feasibility of the main methanogenic pathways relevant for crude oil alkane degradation. For example, the free energy change of methanogenesis from $CO_2$ reduction and acetoclastic methanogensis remain negative up to $CH_4$ concentrations of $10^{24}$ and $10^6$ atmospheres respectively (Figure 8). It should, however, be noted that these calculations were conducted under standard conditions with all reactants and products at 1 M concentration (except methane). Interestingly, in the specific case of acetoclastic methanogenesis the effect of end product concentration is profound. At acetate concentrations that are realistic for petroleum reservoirs (micromolar levels), acetoclastic methanogenesis becomes thermodynamically unfavorable at rather modest methane and $CO_2$ concentrations (a few atmospheres; Figure 9). This is possibly a further reason that under some circumstances the MADCOR process of hydrocarbon degradation is channeled through syntrophic acetate oxidation coupled with methanogenic $CO_2$ reduction, rather than by direct acetoclastic methanogenesis (Jones et al., 2008). A further interesting consequence of increasing $CO_2$ partial pressure from geological $CO_2$ sequestration technologies is the observed shift in carbon flow from syntrophic acetate oxidation coupled with methanogenic $CO_2$ reduction, to acetoclastic methanogenesis (Mayumi et al., 2013).

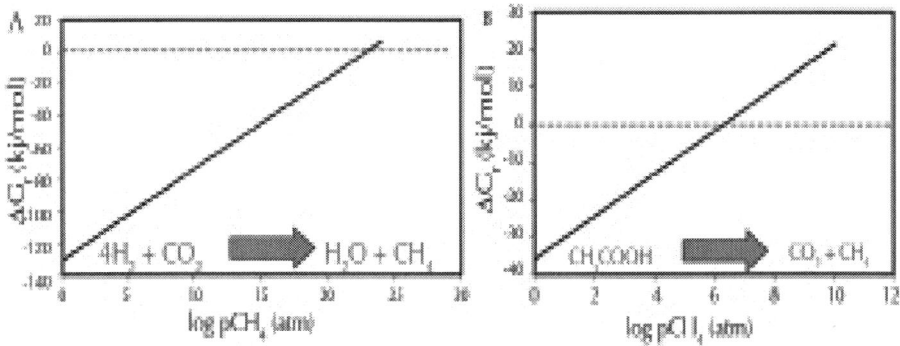

**Figure 8:** The effect of methane partial pressure on the thermodynamic feasibility of (A) methanogenic $CO_2$ reduction and (B) acetoclastic methanogenesis.

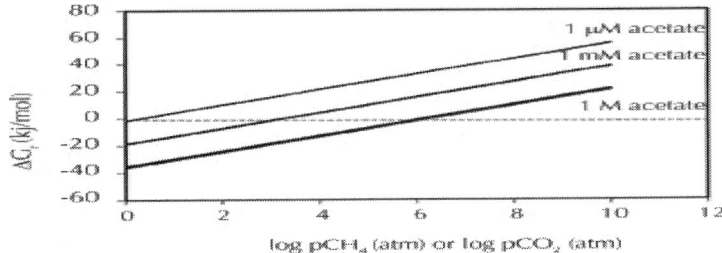

**Figure 9:** The effect of methane or $CO_2$ partial pressure on the thermodynamic feasibility of acetoclastic methanogenesis at different acetate concentrations. At micromolar levels of acetate, acetoclastic methanogenesis becomes thermodynamically unfavourable at modest levels of methane or $CO_2$.

# RATES AND FLUXES—FROM FUNDAMENTAL SCIENCE TO OIL FIELD APPLICATIONS

Biological and biogeochemical inferences about subsurface petroleum biodegradation, have real-world applications. "Kinetic models," of subsurface petroleum biodegradation are now routinely applied in basin modeling studies to assess oil properties prior to drilling expensive exploration or production wells. Based on our assessments of temperature controlled biodegradation rates and zero order biodegradation fluxes (net rates of hydrocarbon destruction, per square meter of oil water contact area within the reservoir), and the development of compositional biodegradation models for crude oil (Larter et al., 2003, 2006), several groups have now developed biodegradation models which consider effects on oil composition (Blumenstein et al., 2008; Haeseler et al., 2010). Larter et al. (2006), calibrated compound class specific or bulk oil degradation flux vs. reservoir temperature profiles using measured compositional gradients in actual oilfields and also used Monte Carlo simulations of large oil data sets from individual basins. The subsurface reservoir biodegradation fluxes (the rate at which petroleum compounds are destroyed in units of kilograms of "hydrocarbons," per square meter of oil water contact area per year), for fresh petroleum in clastic e.g., sandstone, reservoirs are on the order of $10^{-4}$ kg/m$^2$/year at 40°C decreasing with increasing reservoir temperature to a value close to zero, around 80°C. At very low reservoir temperatures (circa 10–20°C), such as seen in the near surface Canadian oil sands, the net degradation fluxes are much less than maximum values. Of course these rates are subject to the other controls e.g., toxicity, nutrients and salinity discussed above.

The degradation flux concept is easily integrated into basin modeling software (which simulates the geological development of basins, the maturation of source rocks, migration of oil and subsequent alteration processes of the oil within the reservoir) as it directly relates total degradation flux to evolving oil water contact area as oilfields develop. The biodegradation flux approach of Larter et al. (2006) enhanced by the incorporation of water leg size and oil compositional effect as controls on subsurface oil biodegradation rates (Adams et al., 2006; Adams, 2008) is now widely utilized in industry. These models indicate that oil charge rates (the volumetric rate of oil charge to a filling reservoir), is a primary control on the quality of oil in a biodegrading oil reservoir. Larter et al. (2003), showed that in fact, the rates of charge and hydrocarbon destruction through biodegradation in heavy oil reservoirs, are commonly of a similar order of magnitude. Thus, small changes in charge rate can greatly affect the net levels of biodegradation seen in the oil field as fresh oil mixes with oil that is already biodegrading.

# MICROBIAL PROCESSES AND IN-RESERVOIR PETROLEUM BIODEGRADATION

## Distinct Patterns in Microbial Communities Enriched in Oil-degrading, Methanogenic, and Sulfate-reducing Conditions

Analysis of microbial communities from oil-degrading, methanogenic and sulfate-reducing microcosms inoculated with the same sediment and identical in every way except sulfate concentration has been reported recently by Gray et al. (2011) and Sherry et al. (2012). Comparison of denaturing gradient gel electrophoresis data from these oil-degrading systems and microcosms incubated under iron-reducing conditions shows that the bacterial communities enriched under different electron accepting conditions are distinct (Figure 10). *Deltaproteobacteria* from the*Syntrophaceae*, most closely related to bacteria from the genus *Smithella* were predominant in the methanogenic oil-degrading systems. This group is often detected at high frequency in methanogenic hydrocarbon-degrading systems (Dojka et al., 1998; Zengler et al., 1999; Bakermans and Madsen, 2002; Kasai et al., 2005; Allen et al., 2007; Shimizu et al., 2007; Ramos-Padrón et al., 2011;Siddique et al., 2011) and is implicated as the primary alkane utilizer in methanogenic crude oil alkane-degrading systems, at least in near surface oil contaminated sites (Gray et al., 2011) if not petroleum reservoirs (see Section Microbial Communities in Petroleum Reservoirs).

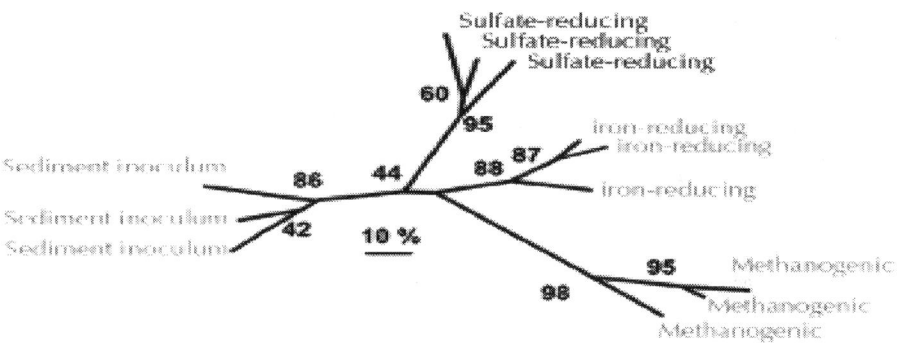

**Figure 10:** Comparative analysis of bacterial community DGGE profiles from crude oil degrading microcosms incubated under different electron-accepting conditions and communities in the sediment used to inoculate the sediments. Distinct communities were selected under different electron accepting conditions. The numbers at nodes represent the percentage of trees in which the group to the right of the node was recovered in datasets subject to bootstrap resampling (100 replicates) and gives an indication of the confidence that can be placed in the groups recovered. The scale represents 10% difference in the community composition based on Dice similarity of pairwise comparisons of DGGE profiles.

Contrary to expectation, known sulfate-reducing taxa were not enriched in oil degrading sulfate-reducing microcosms. During the period of most extensive sulfate-reduction coupled to crude oil alkane degradation, *Gammaproteobacteria* and *Firmicutes* were prevalent (Sherry et al., 2012) and represented by members of the genus *Marinobacterium* and members of the family *Peptostreptococcaceae* respectively. These organisms have been associated with hydrocarbon impacted environments in the past. Though considered aerobic, some *Marinobacterium* strains are known to be facultatively anaerobic (Kim et al., 2007). Moreover *Marinobacterium* has been identified in an oil-water separator tank from a Dutch oil field (van der Kraan et al., 2010) and a saline petroleum reservoir (Yuehui et al., 2008). *Firmicutes* have also been associated with anaerobic crude oil degrading systems (Gieg et al., 2008, 2010; Wang et al., 2011). Thiosulfate-reducing bacteria (TRB) from the families *Clostridiaceae* and *Peptostreptococcaceae* have been isolated from production waters in an onshore oil field in North-Eastern India, (Agrawal et al., 2010). It is therefore possible that the *Firmcutes* found at high frequency in the sulfate-reducing oil degrading microcosm are TRB rather than SRB. About 68–78% of the $HS^-$ generated by sulfate reduction in sediments is re-oxidized to thiosulfate (Jorgensen, 1990) and endogenous generation of thiosulfate in the microcosms could potentially lead to the enrichment of thiosulfate-reducing hydrocarbon degraders. This is consistent with the observation that the sulfate- (and thiosulfate-) reducing archaeon *A. fulgidus* VC-16 degrades alkanes at 70°C with thiosulfate as an electron acceptor (Khelifi et al., 2014). In relation to the occurrence of thiosulfate reducers in petroleum reservoirs it is pertinent that many fermentative bacteria that have been identified in petroleum reservoirs are also capable of thiosulfate-reduction (Magot et al., 2000).

Nevertheless, re-oxidation of sulfide to thiosulfate still relies on sulfate reducers to produce sulfide in the first instance and their lack of strong enrichment may therefore result from oil being degraded syntrophically by consortia of hydrocarbon fermenting organisms with sulfate reducers simply consuming fermentation products. The partition of energy between the fermenters and sulfate-reducing syntrophs in such a system is dependent on *in situ* conditions such as sulfate concentration and accumulation of hydrogen and acetate for example as fermentation end products could result in preferential enrichment of the alkane fermenters relative to the sulfate-reducing terminal oxidizers in the system, leading to an apparent lack of enrichment of classical sulfate-reducers as has been observed (Sherry et al., 2012). We have conducted a thermodynamic analysis illustrating how energy partitioning between syntrophic hexadecane-degrading bacteria and sulfate-reducing terminal oxidizers changes with time as sulfate becomes depleted. As sulfate is depleted, sulfide and bicarbonate increase with hydrocarbon degradation. At acetate concentrations typical of sulfate-reducing sediments, the energy yield for alkane fermentation exceeds the energy available from hydrogen or acetate oxidation with sulfate (Figure 11) potentially leading to a situation that sulfate-reducing taxa are not predominant under conditions where extensive hydrocarbon degradation has occurred under sulfate-reducing conditions.

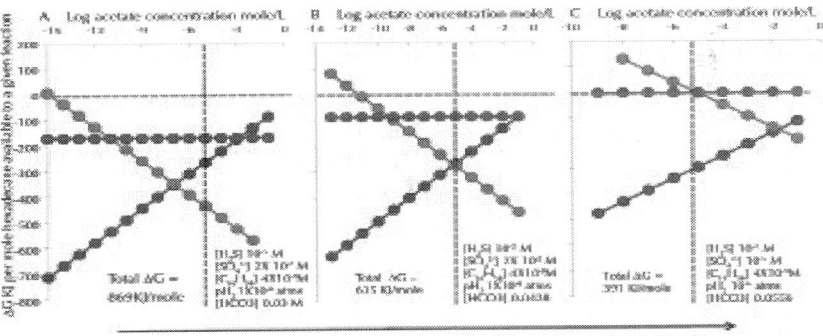

**Figure 11:** The effect on thermodynamic yields of progressive changes in sulfate, sulfide and bicarbonate concentration during hexadecane degradation coupled to sulfate reduction ($C_{16}H_{34}$ + 12.25$SO_4^{2-}$ + 8.5$H^+$ → 16$HCO_3^-$ + 12.25$H_2S$ + $H_2O$). For each set of sulfate, sulfide and bicarbonate conditions (A–C) energy yields were calculated for a range of acetate concentrations and a fixed hydrogen partial pressure for a hypothetical microbial consortium comprising: (red circles and line) hexadecane fermentation to hydrogen and acetate ($C_{16}H_{34}$ + 16$H_2O$ → 8$CH_3COO^-$ + 8$H^+$ + 17$H_2$), (blue circles and line) hydrogen oxidation coupled to sulfate reduction (17$H_2$ + 8.5$H^+$ + 4.25$SO_4^{2-}$ > 4.25$H_2S$ + 17$H_2O$), (green circles and line) Acetate oxidation coupled to sulfate reduction (8$CH_3COO^-$ + 8$H^+$ + 8$SO_4^{2-}$ > 16$HCO_3^-$ + 8$H_2S$). Calculations were performed using the concentrations indicated in each panel. All other conditions used in calculations were fixed e.g., 20°C, pH7 and fixed hexadecane (aqueous solubility) and hydrogen partial pressures p$H_2$ ($10^{-6}$ atms). The vertical dashed line represents a typical threshold level for acetate in sulfate reducing systems.

# Differences in Hydrocarbons Degraded, Mechanisms and Metabolites under Methanogenic and Sulfate-reducing Conditions

Consistent with the prevalence of different bacterial populations in oil degrading systems under methanogenic and sulfate-reducing conditions, differences in the degradation of hydrocarbons e.g., biphenyl isomers (Figure 12) amongst other oil components, are seen under sulfate-reducing and methanogenic conditions (Townsend et al., 2003; Jones et al., 2008). This observation coupled with oil geochemistry from biodegraded reservoirs has provided one piece of evidence that in-reservoir oil degradation is driven by methanogenesis (Jones et al., 2008; Figure 12). Interestingly, in oil-degrading, sulfate-reducing laboratory incubations, considerable accumulation of alkylsuccinates, key products of fumarate addition reactions, was observed. By contrast during methanogenic oil degradation these metabolites were present only at levels similar to control incubations without oil, and levels detected in the sediment used to inoculate the microcosms (Figure 13; Aitken et al., 2013). Similar observations have been reported for other methanogenic oil degrading systems (Gieg et al., 2010; Zhou et al., 2012) suggesting that under methanogenic conditions crude oil alkanes may be activated by a different mechanism. Alternatively the kinetics of alkylsuccinate consumption in the methanogenic alkane-degrading systems may be more rapid leading to lower concentrations of these key intermediates. Interestingly, it has been shown that methanogenic toluene degradation is most likely initiated by fumarate addition, though in this case the corresponding benzylsuccinates accumulated transiently (Beller and Edwards, 2000; Fowler et al., 2012). Quantification of alkylsuccinate synthase genes (assA) also demonstrated that these only increased in abundance during sulfate-driven crude oil alkane degradation and not in methanogenic oil degrading systems. This suggested that organisms harboring assA genes were not enriched under methanogenic conditions. An alternative mechanism for alkane activation under methanogenic conditions was supported by recent metagenomic analysis of a methanogenic hexadecane-degrading enrichment culture (Embree et al., 2013). Embree et al. (2013) did not identify assA genes in the metagenome data, however reanalysis of the data demonstrated that assA genes were present, but they had been annotated as pyruvate-formate lyase genes by the automated annotation pipeline used in the initial analysis of the metagenome (Tan et al., 2014). Not only this, but reanalysis of the metatranscriptome data from the study of Embree et al. (2013) showed that the gene was up-regulated during alkane degradation (Tan et al., 2014). Moreover, a defined mixed culture of *Desulfatibacillum alkenevorans*AK-01, which activates alkanes via fumarate addition, and *Methanospirillum hungatei* JF-1 converts alkanes to methane (Callaghan et al., 2012). It is thus clear that the landscape of anaerobic alkane activation is not straightforward and it may be difficult to associate alkane degradation under particular electron-accepting conditions, with a specific alkane activation mechanism.

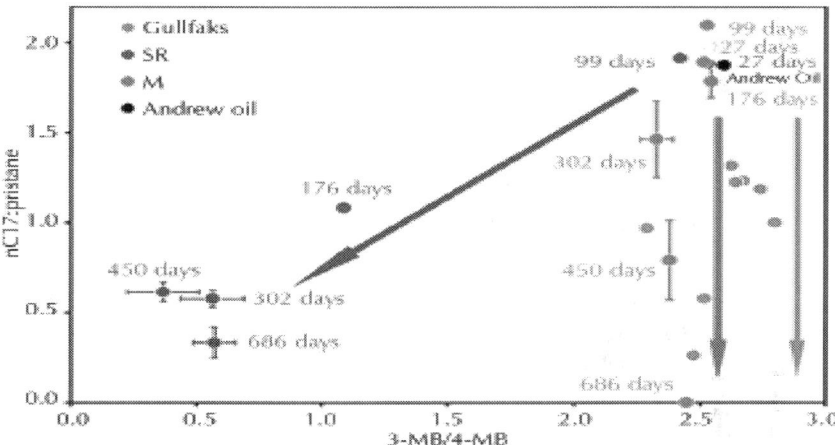

**Figure 12:** Plot of the ratios of n-heptadecane to pristane against the 4-methylbiphenyl to 3-methylbiphenyl ratio from North Sea crude oil degraded in laboratory anaerobic microcosm experiments under sulfate-reducing (SR, blue) and methanogenic (M, red) conditions and field-degraded oils from the North Sea Gullfaks field (Gullfaks, gray). Error bars for the peak ratios are ±1 *SE* (n = 3). The laboratory methanogenic microcosm data and the field data plot along the same biodegradation trajectory with n-alkane degradation but no apparent aromatic hydrocarbon degradation, while the sulfate-reducing microcosm data show the concomitant degradation of n-alkanes and aromatic hydrocarbons under different electron accepting conditions.

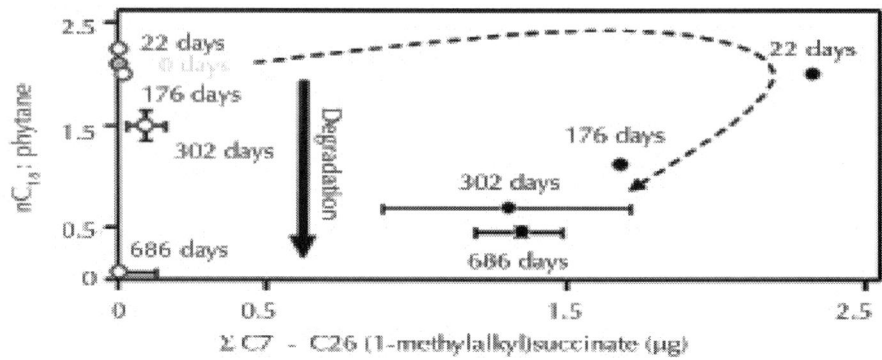

**Figure 13:** Changes in nC18 to phytane ratio with total C7 to C26 (1-methylalkyl) succinates (µg) in methanogenic microcosms (open circles) and sulfate-reducing microcosms (filled circles) over 686 days of anaerobic hydrocarbon degradation. Error bars, where shown, are ± one standard error of replicate microcosms where $n = 3$. The dotted arrow indicates the temporal changes in alkylsuccinate concentration under sulfate-reducing conditions.

Hydrocarbon addition to fumarate which involves formation of substituted succinates as catabolic intermediates is the best characterized mechanism for activation of hydrocarbons in the absence of oxygen. This mechanism has been observed under a wide range of electron accepting conditions and represents a central paradigm in anaerobic hydrocarbon degradation. However, alternative pathways have been proposed. For instance sub-terminal carboxylation of alkanes at C-3, with the formation of 2-ethyl fatty acids with subsequent removal of the two terminal carbons has been suggested (Aeckersberg et al., 1998; So et al., 2003; Boll and Heider, 2010). This results in formation of C-odd fatty acids from C-even alkanes (and vice versa) by the sulfate-reducing strain *Desulfococcus oleovorans* Hxd3. Interestingly So et al. (2003) were unable to detect 2-ethyl fatty acid intermediates and likewise they were not detected in the acid fraction from the crude-oil degrading methanogenic microcosms investigated by Aitken et al. (2013). In addition, there is a yet undetermined mechanism for the formation of C-even cellular fatty acids from C-even alkanes by a denitrifying bacterium *Pseudomonas balearica*. This organism utilizes C15 -C18 alkanes as substrate but does not appear to use fumarate addition for initial alkane activation (Grossi et al., 2008).

Anaerobic hydroxylation of an alkyl carbon in ethylbenzene by a molybdenum-containing enzyme has also been observed (Johnson et al., 2001). Hydroxylation of alkanes is considered less likely due to the higher C-H bond dissociation energy of alkanes (400 kJ/mol) compared to the C-H bond dissociation energy of C-2 of the alkyl side-chain of ethylbenzene (355 kJ/mol) (Boll and Heider, 2010). Alkane C-H bond dissociation energy reduces slightly with alkane chain length and is slightly lower for C-2 of n-alkanes and secondary or tertiary carbon atoms, but such reactions have only been considered feasible when a high potential electron acceptor such as nitrate or FeIII is used. This is consistent with the observation that nitrate- and iron-reducers are capable of hydroxylation of the methyl group of cresol, but typically sulfate-reducers degrade cresol by fumarate addition (Boll and Heider, 2010). Nevertheless, there is no *a priori* reason that hydroxylation of an aliphatic carbon with a C-H bond dissociation energy greater than 355 kJ/mol is not possible and it now seems that alkane activation by *Desulfococcus oleovorans* may not proceed by initial carboxylation at C-3 of the alkane but instead by hydroxylation at C-2 followed by oxidation of the hydroxyl group and carboxylation at C-3 of the methylalkyl ketone (Callaghan, 2013).

# MICROBIAL COMMUNITIES IN PETROLEUM RESERVOIRS

Most data on microbial communities in petroleum reservoirs come from produced waters and there are almost no data on sediments recovered from petroleum reservoirs by coring (Magot, 2005; Head et al., 2010). The nature of produced waters, and issues as to whether or not water injection for secondary recovery has been practiced leads to difficulties in assigning the provenance of organisms cultured from petroleum reservoirs (Stetter et al., 1993; Magot, 2005). A substantial proportion of the organisms identified in culture independent analyses of produced waters from petroleum reservoirs are therefore considered to be non-indigenous organisms that have grown in the production infrastructure (Orphan et al., 2000; Grabowski et al., 2005). To date there have been few studies which have focused specifically on the microbial communities present in biodegraded petroleum reservoirs (Grabowski et al., 2005; Sette et al., 2007; Dahle et al., 2008; de Oliveira et al., 2008; Hallmann et al., 2008; Hubert et al., 2012; Kobayashi et al., 2012). In some of these there were apparently no differences in the communities present in degraded and non-degraded reservoirs (Sette et al., 2007; de Oliveira et al., 2008) though the relative abundance of taxa detected in 16S rRNA gene clone libraries did differ in one study (Sette et al., 2007).

A recent broad survey of microbial community data from a range of oil and hydrocarbon-impacted anoxic environments has demonstrated that the group of organisms found most frequently, and at the highest relative abundance is the *Firmicutes* followed by the *Gamma-, Delta-* and*Epsilonproteobacteria* (Gray et al., 2010; Figure 14). This survey included data from systems that contained both biodegraded and non-biodegraded oil and also some examples of enrichment cultures and microcosm studies where active hydrocarbon degradation was occurring. The analysis indicated that *Smithella* spp. may play a central role in methanogenic crude oil alkane degradation in anoxic environments lacking any alternative electron acceptors (Head et al., 2010; Gray et al., 2011; Cheng et al., 2013a). However, although a number of surveys report the detection of*Syntrophaceae* related to *Smithella* and *Syntrophus* in petroleum reservoir systems (Gray et al., 2011 and reference therein) the majority of studies relevant to biodegraded oil fields indicate that they are detected at low frequency (Gieg et al., 2008; Wang et al., 2011; Kryachko et al., 2012;Mbadinga et al., 2012). Crude oil and hydrocarbon-degrading enrichments inoculated with oil field waters reveal that other organisms may provide the key function of alkane fermentation, especially in high temperature systems (Gieg et al., 2010; da Cruz et al., 2011; Wang et al., 2011; Mbadinga et al., 2012). In the majority of these cases *Firmicutes*, predominantly *Clostridia* were detected at highest frequency (19–71% of 16S rRNA gene clones; (Gieg et al., 2010; Wang et al., 2011; Mbadinga et al., 2012). In one case *Bacillus* and *Acinetobacter* were predominant (da Cruz et al., 2011) although the medium used in this study contained 8 mg of yeast extract in each 40 ml microcosm as well as 30 mg of oil, of which only a small fraction would have been biodegraded over the time course of the experiments (da Cruz et al., 2011). Control incubations with no oil were also run in this study, however the composition of the microbial communities in the control incubations was not reported (da Cruz et al., 2011).

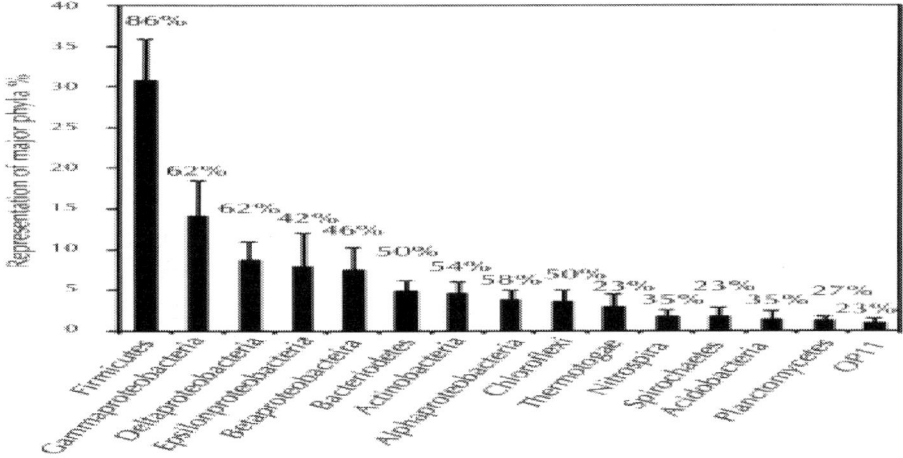

**Figure 14:** Frequency distribution of 16S rRNA sequences (classified into major phylogenetic groups) recovered in clone libraries from hydrocarbon impacted environments. Bars correspond to average percent representation of major phyla (1× *SE*) based on a survey of 26 bacterial clone libraries. Values shown above the columns indicate the percentage of studies in which the phylum was identified. Modified from Gray et al. (2010).

The significance of *Firmicutes*, especially those related to *Clostridia*, in anaerobic hydrocarbon degradation associated with petroleum reservoirs seems increasingly likely. *Firmicutes* identified in methanogenic oil degrading systems include organisms related to known syntrophic acetate oxidizing bacteria such as *Thermacetogenium* and *Moorella* (Gieg et al., 2010; Mbadinga et al., 2012). The presence of syntrophic acetate oxidation to $H_2$ and $CO_2$ was linked to a shift in methanogenic communities from acetoclastic to hydrogenotrophic methanogens with an increasing number of transfers of a methanogenic hexadecane-degrading enrichment, inoculated with water from the Shengli oilfield, China (Cheng et al., 2013c). This is particularly interesting given the prevalence of hydrogenotrophic methanogenesis by $CO_2$ reduction in petroleum reservoirs (Head et al., 2010) and the fact that syntrophic acetate oxidation has been implicated as a central process in methanogenic crude oil biodegradation in some systems (Jones et al., 2008; Gray et al., 2011). The significance of *Firmicutes* is also suggested by the increase in concentration of anteiso branched phospholipid fatty acids detected in biodegraded oils (Hallmann et al., 2008). In higher temperature enrichments (55 to 74°C; Gieg et al., 2010; Mbadinga et al., 2012) thermophilic taxa (e.g.,*Thermodesulfobiaceae* within the *Clostridia*, and *Thermotogae*) were also enriched opening up the possibility that a relatively broad range of organisms may have the capacity for anaerobic degradation of different crude oil components. Interestingly, the organisms selected in 55°C enrichment from an Alaskan oil reservoir were also detected in produced waters from the same reservoir (Duncan et al., 2009; Gieg et al., 2010). The microbial communities enriched in high temperature oil degrading systems appear to be quite different from those observed under mesophilic conditions (Gray et al., 2010). Gieg et al. (2010), for example identified organisms from the *Thermotogales* (closest relative *Thermotoga elfii*), *Deferribacterales* (related to *Flexistipes*spp.), *Synergistales* (related to *Anaerobaculum hydrogeniformans* OS1) and*Thermoanaerobacterales* (related to *Thermacetogenium phaeum*, a syntrophic acetate oxidizer from the *Firmicutes*). A high temperature (55°C) methanogenic enrichment culture which degraded hexadecane has also been found to be dominated by a novel deep branching bacterial group related to *Thermotoga* spp. and has been designated the Shengli Cluster (Cheng et al., 2013b). It is interesting to speculate that these organisms might represent an entirely new group of thermophilic hydrocarbon-degrading syntrophs.

# MICROBIAL COMMUNITIES IN HEAVILY BIODEGRADED RESERVOIRS

When one examines the composition of microbial communities from analysis of produced waters from the most heavily biodegraded oil reservoirs, the picture that emerges is somewhat different. Microbial communities associated with waters from heavy oil fields in the Western Canadian Sedimentary Basin and Oil Sands reservoirs appear to harbor communities that are dominated by*Epsilonproteobacteria* (Grabowski et al., 2005; Hubert et al., 2012). It has been suggested that this dominance is artifactual caused by selectivity of the primers used for 16S rRNA gene analysis (Grabowski et al., 2005), however, more recent work points to the genuine dominance of *Epsilonproteobacteria* in these systems (Hubert et al., 2012). *Epsilonproteobacteria* were also detected as abundant members of coal bed microbial communities (US Patent 2010/0047793 A1, 2010). The fact that the petroleum in heavy oil reservoirs has low levels of saturated hydrocarbons like *n*-alkanes and contain higher levels of aromatic hydrocarbons and more polar aromatic components such as benzothiophenes, might explain why the microbial communities in biodegraded heavy oil reservoirs appear to be distinct. At present the specific role of the resident microbiota in heavy oil biodegradation is unknown, For instance, many *Epsilonproteobacteria* are putative chemoautotrophic, nitrate-reducing, sulfide oxidizers and some may be responsible for sulfide oxidation in sour petroleum reservoirs (Gevertz et al., 2000; Hubert, 2010). However, they may be autotrophic sulfide oxidizing autotrophs and heterotrophs utilizing a range of electron acceptors (Campbell et al., 2006). *Sulfuricurvum kujiense* is regarded as a sulfide oxidizing chemoautotroph but the type strain YK-1 can obtain energy for growth by oxidizing reduced, probably organic, sulfur in crude oil (Kodama and Watanabe, 2003, 2004). This could explain the abundance of *Sulfuricurvum* in highly biodegraded oil reservoirs such as the Canadian oil sand reservoirs, where the oil has a relatively high sulfur content (Strausz and Lown, 2003). Moreover the oil-water transition zone between oil sands reservoirs and underlying aquifers can exhibit gradients of a range of aromatic hydrocarbons and sulfur heterocycles such as dibenzothiophenes and such compositional gradients are indicative of active biodegradation in transition zones in oil sands (Head et al., 2003; Hubert et al., 2012).

Oxidation of reduced sulfur compounds *in situ* may be driven by nitrate reduction (Hubert, 2010) or reduction of metal species (Aller and Rude, 1988). Indeed some *Arcobacter* and *Sulfurospirillum* spp. are capable of iron or manganese reduction (Thamdrup et al., 2000). *S. kujiense* strain YK-1, like some other *Epsilonproteobacteria*, can grow under microaerophilic conditions (Kodama and Watanabe, 2003, 2004) and use oxygen as an electron acceptor (Campbell et al., 2006). It is therefore also possible that small amounts of oxygen introduced from meteoric water into aquifers associated with shallow heavily degraded oil sands reservoirs, could provide an electron acceptor for sulfide or organic sulfur oxidation.

Some studies suggest *Epsilonproteobacteria* might play a role in syntrophic anaerobic hydrocarbon-degrading communities. Assimilation of $^{13}C$ into nucleic acids from an epsilonproteobacterium during syntrophic degradation of $^{13}C$-labeled benzene in sulfate-reducing enrichment cultures has been observed, and in fact the highest degree of $^{13}C$ enrichment was observed for the *Epsilonproteobacteria* (Herrmann et al., 2010). It is possible, therefore, that this organism may contribute to benzene fermentation to acetate, $H_2$ and $CO_2$ that are subsequently used by the sulfate-reducers in the system. *Sulfurospirillum* spp. are metabolically versatile and can ferment a wide variety of organic compounds (Luijten et al., 2003) and were among the fermentative heterotrophs isolated from a subsurface coal deposit (Fry et al., 2009). *Sulfurospirillum* spp. were also isolated from the Pelican Lake oil reservoir in northern Alberta (Grabowski et al., 2005). *Arcobacter* spp. utilize acetate under anoxic conditions (Thamdrup et al., 2000; Fedorovich et al., 2009; Webster et al., 2010) and several *Epsilonproteobacteria* can use $H_2$ as an electron donor (e.g., Gevertz et al., 2000; Kodama et al., 2007), including an acetogenic *Arcobacter* abundant in the Pelican Lake reservoir bacterial community (Grabowski et al., 2005).

# PUTATIVELY AEROBIC ORGANISMS AND FUNCTIONAL GENES IN PETROLEUM RESERVOIRS

Increasingly in the literature on hydrocarbon-containing subsurface reservoirs there are reports of organisms conventionally considered to be aerobic (or in some cases facultative) heterotrophs e.g., *Bacillus* spp., *Acinetobacter* spp. and *Pseudomonas* spp., particularly in heavy, biodegraded oil reservoirs such as the Alberta oil sands (Orphan et al., 2000; da Cruz et al., 2011; Li et al., 2012; Zhang et al., 2012; An et al., 2013; Meslé et al., 2013). This raises questions about the anaerobic/methanogenic origin of heavy oil that has prevailed for a number of years (Head et al., 2003). What then, does the occurrence of these apparently aerobic organisms indicate about the biodegraded petroleum systems where they occur?

There are a number of possible explanations for these observations.

- Samples were exposed to oxygen during sampling, transport and storage allowing growth of aerobic organisms initially present at low abundance.

- Oxygen was supplied externally to the reservoir by meteoric water.

- A "cryptic" aerobic community that uses *in situ* generated oxygen may be present in some petroleum systems.

- The aerobes detected are in fact capable of anaerobic metabolism.

The first of these, oxygen exposure and outgrowth of populations of aerobes initially present at low abundance, is certainly a possibility given the technical difficulties in obtaining high quality samples from operational petroleum systems. However, one might expect, for example, that a suite of samples from the same core would be equally likely to be exposed to oxygen and therefore all samples should contain this aerobic microbial community signal. This does not seem to be the case and communities exhibiting primarily aerobic signatures have been found adjacent to samples harboring communities with a predominantly anaerobic character (An et al., 2013).

If aerobic organisms are indeed residents of petroleum reservoirs, where might oxygen come from in deep subsurface sediments? Oxygen could be delivered to oil sands reservoirs in meteoric water, as envisaged in the classical model of in-reservoir petroleum biodegradation (Palmer, 1993). This ingress might be facilitated by the relatively shallow depth of reservoirs e.g., the Western Canada Sedimentary Basin, or by faults and fractures providing preferential migration paths for water, delivering oxygen without consumption by microorganisms or reaction with minerals. Repeated glaciation-deglaciation cycles in the oil sands area, over the last 20,000 years, could also have potentially delivered cold-oxygen containing waters to shallower reservoirs and this would be somewhat facilitated by isostatic rebound, pulling water into the sediment as ice sheet loading decreased, in effect, transporting water by "isostatic pumping."

A second possible reason for aerobes in deep subsurface sediments is the occurrence of a "cryptic" aerobic community using oxygen generated *in situ*. Mechanisms are known, or proposed, whereby oxygen may be generated microbially by dismutation of chlorate, nitrate or nitrogen oxides (Achenbach et al., 2006; Ettwig et al., 2012) however these species are not typically found naturally in petroleum reservoirs, though they may be present in reservoirs subject to e.g., nitrate injection for souring control (Hubert, 2010). There is a considerable literature on radiolysis of water generating hydrogen to fuel the deep biosphere (Lin et al., 2005). However, much less consideration has been given to the oxygen species that are also generated by radiolytic splitting of water (Bjergbakke et al., 1989; Draganic, 1991). Interestingly, the obligate aerobe *Deinoccocus radiodurans* only grows anaerobically if the culture is exposed to ionizing radiation presumably making use of oxygen generated from radiolysis of water. Although oxygen generation *in situ* has the capacity to support an aerobic microbial community, the amount generated may not be sufficient to support the degree of oil degradation required to produce heavy oil such as found in the Western Canada sedimentary basis.

If one assumes a rate of water radiolysis of $2 \times 10^{-8}$ nanomoles/$dm^3$/s (based on a range of 1.5 to $4.5 \times 10^{-8}$ nanomoles/$dm^3$/s from data in Lin et al., 2005) then this could generate $H_2$ at rate of $2 \times 10^{-8}$ nanomoles/$dm^3$/s and $O_2$ at a rate of a rate of $1 \times 10^{-8}$ nanomoles/$dm^3$/s. Thus, for 1 $m^3$ of sediment, oxygen could be generated at $10^{-5}$ nanomoles/s or $3 \times 10^{-7}$ moles/y.

Based on oil compositional gradients, areal oil degradation fluxes at the OWTZ of reservoirs containing heavy oil have been estimated at around $10^{-5}$ to $10^{-6}$ $kg/m^2$/y (Larter et al., 2003). Given that oil is about 85% carbon this results in a carbon flux of $8.50 \times 10^{-3}$ to $8.50 \times 10^{-4}$ g $C/m^2$/y or $7.1 \times 10^{-4}$ moles $C/m^2$/y. On this basis we can estimate a fractional rate for aerobic oil degradation ($F_{aerobic}$ = aerobic degradation/total degradation). To convert volumetric rates to fluxes we can assume that a given depth of water aquifer, underlying the OWTZ, contributes to the oxygen fed to it. Thus, for 1 m of water leg contributing oxygen, $F_{aerobic} = 3 \times 10^{-7}/7.1 \times 10^{-4} = 3 \times 10^{-4}$ or 0.042%. For 100 m of water leg $F_{aerobic} = 3 \times 10^{-5}/7.1 \times 10^{-4} = 3 \times 10^{-2}$ or 4.2% of the total degradation flux. These are very crude calculations, but the values obtained are likely to be overestimates and suggest complete aerobic oxidation of hydrocarbons by this mechanism is not significant. Radioloysis of water generates a number of different oxygen species not just $O_2$ (Bjergbakke et al., 1989; Draganic, 1991) and these are more reactive than oxygen and probably not available to microbial communities. It is also unlikely that all of the oxygen generated from the aquifer underlying a reservoir would reach the

OWTZ reinforcing the conclusion that radiolysis of water could only make a small contribution to aerobic oil degradation in the subsurface. It is interesting to speculate, however, that trace levels of oxygen are not used to completely mineralize organic compounds, but instead for an initial activation of the hydrocarbon molecules, with subsequent anaerobic metabolism. Such dysaerobic activation and anaerobic metabolism (DAAM), could then be much more significant, but still a minor contributor to net biodegradation rates. Moreover, during biodegradation of light oil, high levels of labile substrates are present in the oil and anaerobes might effectively compete with aerobes which are limited by very low oxygen levels. However, in heavy oil, electron donors and carbon sources may be more limiting and aerobes will be able to compete more effectively even if oxygen levels are very low which may lead to a situation where aerobes might dominate the habitat due to the higher growth yields that can be achieved with oxygen as an electron acceptor. It should however also be considered that maintenance energy requirements for aerobes are also typically higher than for anaerobes and thus at very low oxygen concentrations they may not achieve high growth yields.

The bulk, molecular and gas geochemistry of heavy oilfields is also a major bounding constraint, with the gases associated with heavy oilfields showing very dominant methane and minor $CO_2$ content and carbon isotopic signals inconsistent with large-scale aerobic processes. Thus, in the Albertan heavy oil fields (Adams, 2008; Adams et al., 2013a), the carbon isotopic signatures of associated carbon dioxide in equivalent reservoirs are much heavier ( $^{13}C$ $CO_2$ of −10 to +20 per mil typically), than that expected by direct aerobic oxidation of source crude oils. Oils sourced from the Exshaw and Gordondale Fm. in this area, have whole oil $^{13}C$ signatures of around −30 to −31 per mil (Adams et al., 2013a), and they would be expected to produce $CO_2$ as the dominant gaseous end product, and this would have a similar carbon isotopic signature to the source oils, if formation of heavy oil in the oil sands was predominantly a product of aerobic biodegradation. Thus, the gas geochemistry of the heavy oils and oil sands is not consistent with aerobic biodegradation being a very significant formative process. The dominant methane, subordinate $CO_2$ and often very heavy carbon isotopic signatures of the $CO_2$ observed in heavy oils and oil sands reservoirs, is consistent with anaerobic and specifically methanogenic processes (Jones et al., 2008).

The final possibility is that putative aerobes observed in petroleum reservoirs and other deep subsurface sediment environments, are in fact facultative organisms capable of anaerobic metabolism. *Pseudomonas* spp. for example, classically thought of as catabolically versatile aerobes or facultative aerobes that utilize nitrate and other oxidized nitrogen species as alternative electron acceptors to oxygen, are probably more cosmopolitan in their use of electron acceptors than commonly considered. At least one *Pseudomonas* sp. (*Pseudomonas aeruginosa* strain KRP1) is known to use solid phase anodes in a microbial fuel cell (MFC) as an electron acceptor and it achieves this through the mediation of phenazine electron shuttles that deliver electrons from the cell to the anode of the MFC (Rabaey et al., 2005).

Potentially putative aerobes may even be able to grow as syntrophs in partnership with methanogens. This tantalizing possibility has recently been raised by a study of a methanogenic, crude oil-degrading enrichment culture inoculated with oil reservoir production water (Berdugo-Clavijo and Gieg, 2014). In this study, a methanogenic oil degrading consortium, dominated by *Smithella* sp., a putative alkane fermenting organism, and a range of acetoclastic and $CO_2$-reducing methanogens, was inoculated into anoxic sand columns containing residual oil, and incubated in an anoxic chamber for over 300 days. The sand columns actively degraded crude oil hydrocarbons and generated methane. When the microbial communities in the sand columns were analyzed, the composition of the microbial communities had changed considerably with $CO_2$-reducing methanogens predominating in the community. However, the most intriguing observation was that the most highly represented bacterial taxon was a *Pseudomonas* sp. (Berdugo-Clavijo and Gieg, 2014). While the authors were measured in their interpretation that this *Pseudomonas* sp. might represent the syntrophic hydrocarbon-degrading partner of the methanogens in the culture, the results do raise the intriguing possibility that some *Pseudomonas* spp. have evolved to occupy a niche whereby they are active under highly reducing conditions. Their detection in an increasing range of anoxic environments may relate to their capacity to thrive in these environments rather than representing mere aerobic contaminants. This view is strengthened by anaerobic isolation of *Pseudomonas* sp. GZ1 (Guo et al., 2008). GZ1 was isolated using the Hungate method and shown to ferment organic compounds to hydrogen and may therefore have the capacity to grow in a syntrophic partnership with methanogens for example. Thus, we speculate, that at least some of these putative aerobes are likely, putative no more, and are in fact most likely anaerobic partners in the biodegradation process itself. It would appear that the study of anoxic environments exhibiting petroleum biodegradation continues to overturn conventional learnings of classical microbiology.

# GEOLOGICAL LEARNINGS AND PRACTICAL APPLICATIONS

While much of what is presented above is of importance in understanding the fundamental biogeochemistry of heavy oil reservoirs, it is not without practical significance. It is well know that one of the major impacts of biodegradation on petroleum is to increase oil viscosity primarily by destruction of low molecular weight compounds (Adams et al., 2008; Larter et al., 2008). Biodegradation of crude oil driven from a basal biodegradation zone produces vertical viscosity gradients. These vertical and complementary lateral oil viscosity gradients have major impact on oil production strategies and production rates (Larter et al., 2008). In many reservoirs steps or breaks in the compositional and oil viscosity profiles are common, reflecting reservoir compartmentalization through baffles, barriers or even faults (Fustic et al., 2011). Characterization of vertical and lateral viscosity gradients and intra-reservoir baffles and barriers using geochemistry is now a major commercial application in the heavy oil industry as it allows better producing, less viscous zones in the reservoir to be targeted for production (Sereda and James, 2014). Modeling approaches have also been developed to facilitate the identification of production sweet spots (Larter et al., 2003, 2008). Similar effects are seen in more conventional recovery processes, and it can probably be stated that biodegradation ubiquitously produces fluid property gradients in heavy oil reservoirs and these have a major impact on production strategies and efficiencies.

In-reservoir, crude-oil biodegradation is not therefore merely a phenomenon of academic significance, it impacts the occurrence, distribution and properties of the most abundant form of liquid petroleum on planet Earth, as well as being responsible for generation of a significant fraction of the natural gas that we produce today (Milkov, 2011).

# IMPACT ON ENERGY RECOVERY AND CARBON MANAGEMENT

The discovery of MADCOR raises many possibilities for reduced emission processes for energy recovery from heavy oil. This is because methane is quantitatively produced from biodegradation of hydrocarbons, utilizing water as a co-reactant, with molecular hydrogen gas being the principal intermediate (Jones et al., 2008). If the process could be dramatically accelerated by engineering, hydrogen generation and recovery *in situ*, could provide the basis of a truly green energy recovery process (Larter et al., 2012). In oil reservoirs, acetoclastic methanogenesis seems subordinate and most methanogenesis (>80%), results from reduction of $CO_2$ with hydrogen. Dolfing et al. (2008), demonstrated that, for the crucial steps, a low hydrogen partial pressure (<4 Pa/ $4 \times 10^{-5}$ atm) is required for thermodynamic feasibility and this is achieved naturally in oilfields by methanogens using hydrogen for $CO_2$ reduction. So, while it is theoretically feasible to recover hydrogen from biodegrading oilfields, the low hydrogen concentrations present when the process is taking place actively, would render this difficult to engineer as a large scale process. To achieve this would require the development of dynamically managed recovery systems. These have been proposed, but would be a considerable technical challenge (Larter et al., 2012). The prospect that $CO_2$ produced when alkanes are degraded, with the formation of hydrogen, could be selectively sequestered in-reservoir by manipulating formation water properties is also an attractive element of such a process, but this too, on a reservoir scale, would be a significant technical achievement.

Accelerated methane production through nutrient addition to heavy oilfields is feasible today, but with low natural gas prices in North America resulting from recent increases in shale gas recovery, there is currently little business interest in such schemes. However, increased oil production together with gas is being considered as a biologically assisted engineering recovery process following on from conventional oil recovery (Figure 15A). An alternative to terminal methane production could be use of methanotrophs (methane consuming organisms) to convert produced methane to higher molecular weight hydrocarbons, other chemical species or methanol, directly or indirectly. Biological approaches that use methanotrophs to produce long-chain hydrocarbons have been identified as a priority by the United States government (Chu and Marjumbdar, 2012). Thus, a biological: heavy oil to methane to heavier hydrocarbons or other fuels using methanogenic and methanotrophic routes is possible but seems a stretch at the present.

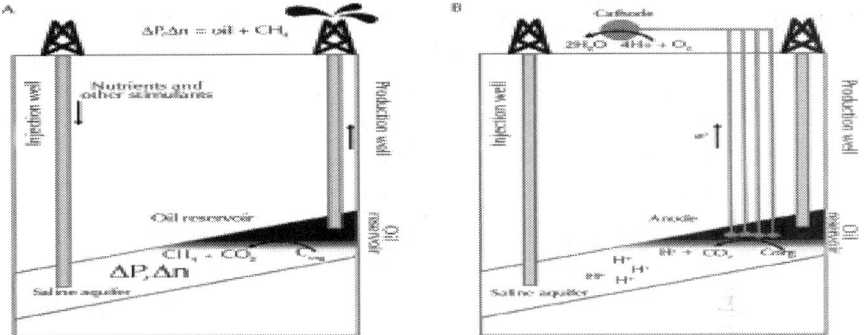

**Figure 15:** Proposed systems for energy recovery from stranded, residual oil in petroleum reservoirs. (A) Enhanced oil and gas recovery by stimulation of methanogenic oil biodegradation and (B) a bioelectrochemical system for the direct recovery of energy, as electricity from residual and heavy oil. Organic carbon ($C_{org}$), Increase in pressure from gas generation ($\Delta P$), decrease in oil viscosity due to gas dissolving in oil ($\Delta\eta$).

Are alternate futures possible therefore? The heavy oil and bitumen resource worldwide, represents a vast chemical resource that could drive various chemical or electrochemical processes *in situ* or *ex situ* that might be used for low emission energy recovery including being a source of hydrogen or electrons via a large scale biolelectrochemical system (Figure 15B). Such a biotechnological application is feasible in principle but implementation on a reservoir or even basin scale is not trivial. A "pure" bioelectrochemical system will also need to contend with issues such as low power output and high internal electrical resistance in a large scale system. However, using the reducing power linked to an accessory power source to synthesize intermediate energy vectors such as in a microbial electrolysis cell (Cheng and Logan, 2007) or to charge batteries (Xie et al., 2013) could be considered as hybrid technologies to overcome some of these fundamental hurdles.

History tells us it is impossible to predict the future over timescales of more than a few months or years. Our best and simplistic estimate of development routes for energy resources over the next few decades is based on both technological readiness and social and political acceptance (Figure 16). Coal and oil use will continue for several decades, but their usage will need to decline substantially before 2050, when it is estimated that industrialized country emissions need to fall 10% below current levels. Today, coal and oil use is socially acceptable as sensible energy pricing and charges for emitted carbon dioxide are politically difficult to implement, but without emissions reduction technology in place, they will likely become socially unacceptable and decline in use. Transition away from traditional fossil fuels will initially involve lower carbon fuel options including greater use of natural gas and increased coupling of carbon capture and storage technologies to the traditional fossil mix, plus inevitable growth in next generation nuclear electricity generation. Carbon capture and storage, and nuclear are currently viewed as only marginally desirable by society, but they will have to grow in the next decade if 2050 emissions targets and a more sustainable energy supply is to be achieved. Solar and renewable energy will also continue to grow in significance, but is unlikely to represent more than 20–30% of the energy base by the same date. However, economically profitable incumbent industries resist change. Carbon dioxide capture and storage probably then provides the first step on the ladder of transition, but carbon neutral fuels (Zeman and Keith, 2008; Graves et al., 2011; Lackner et al., 2012), would also provide an effective strategy whereby carbon dioxide captured from the air is reduced by hydrogen (from low carbon energy sources such as hydroelectric or nuclear electricity), producing the range of transportation fuels that we have today. This techno-economic-finesse would potentially allow the petroleum industry further space for rapid transition, maintaining a hydrocarbon economy while ceasing to produce additional fossil carbon. While much development work still needs to be done, and there is no general agreement on a feasible price for the cost of carbon dioxide capture from the air, it has been suggested that an air capture-based, carbon dioxide to fuel cycle might be a suitable topic for a large strategic focused research program carried out along the lines of the Manhattan or Apollo projects (Broecker, 2013). Where might biotechnology play a role here (beyond plausible Nth generation biofuels, of course)? While current air capture of carbon dioxide schemes involve liquid, solid carbon dioxide sorbent phases (e.g.,Stolaroff et al., 2008), biological capture schemes using novel high pH water chemistries and organisms together with novel bioreactors and fermentative recycling of the generated biomass to methane have been proposed recently (Strous, 2014). This is an area that that has such great potential, albeit with substantial engineering and economic obstacles to overcome, that large growth in this area would be expected. However, the social acceptance of air captured $CO_2$ and carbon neutral fuels remains low (Figure 16), largely because the public is mostly unaware of such possibilities. We suggest that this will be an increasingly important route for investigation as a viable solution to our carbon predicament. In the century of biology, it seems inevitable that biotechnological options for energy production linked to reduction in atmospheric $CO_2$concentrations will be a major research effort.

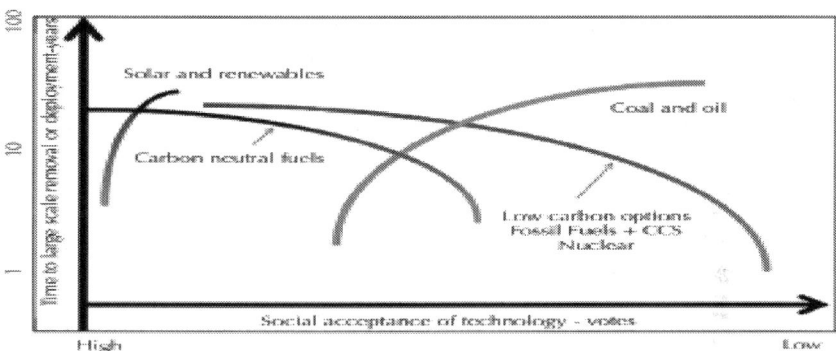

**Figure 16:** The carbon management and energy universe; our best estimate of the likely development routes for different energy resources over the next few decades. The y-axis indicates the timescale to large-scale removal (reduction to 20% of market share, or deployment of a technology type (expansion beyond 20% of market share), and on the x-axis social acceptance and thus deployability of the technology type.

# AUTHOR'S CONTRIBUTIONS

Ian M. Head and Stephen R. Larter conceived and wrote the manuscript, Neil D. Gray edited and revised the paper, provided unpublished data and provided intellectual input to many of the ideas developed in the manuscript.

# ACKNOWLEDGMENTS

Our work on heavy and biodegraded crude oil has been funded over a number of years by the Natural Environment Research Council (NE/ E01657/X1; NE/J024325/1) and three phases of an industrial research consortium (Bacchus I, II, III). Bacchus sponsors involved in one or more of the project phases were Exxon, Norsk-Hydro, Petrobras, Shell, Total-Fina-Elf, Phillips Petroleum, Chevron-Texaco, Japan National Oil Corporation, BP, Conoco, Statoil, Saudi Aramco, Agip/ENI S.P.A., Woodside and Anadarko. Carbon Management Canada and the Genome Alberta, Hydrocarbon Metagenomics Project have also supported our research in this area. A large number of individuals have contributed to this work over many years including, Angela Sherry, Wilfred Röling, Jan Dolfing, Arlene Ditchfield nee Rowan, Martin Jones, Carolyn Aitken, Jennifer Adams, Barry Bennett, Thomas Oldenburg, Haiping Huang, Berni Bowler, Emma Bowen and Casey Hubert. This manuscript includes unpublished data from Carolyn Aitken, Emma J. Bowen, Arlene Rowan and Henry Coombs.

# REFERENCES

1.  Achenbach, L. A., Bender, K. S., Sun, Y., and Coates, J. D. (2006). "The biochemistry and genetics of microbial perchlorate reduction," in *Perchlorate, Environmental Occurrence, Interactions, and Treatment*, eds B. Gu and J. D. Coates (New York, NY: Springer Publishers), 297–310.

2.  Adams, J. J. (2008). *The Impact of Geological and Microbiological Processes on Oil Composition and Fluid Property Variations in Heavy Oil and Bitumen Reservoirs*. Ph.D. thesis, University of Calgary.

3.  Adams, J. J., Jiang, C., Bennett, B., Huang, H. P., Oldenburg, T. B. P., Noke, K., et al. (2008). "Viscosity determination of heavy oil and bitumen. Cautions and solutions," in *World Heavy Oil*

*Conference Edmonton Paper* (Calgary, AB: DMG World Media), 2008–443.

4.    Adams, J. J., Riediger, C., Fowler, M., and Larter, S. R. (2006). Thermal controls on biodegradation around the Peace River tar sands: Paleo-pasteurization to the west. *J. Geochem. Explor.* 89, 1–4. doi: 10.1016/j.gexplo.2005.11.004.

5.    Adams, J. J., Rostron, B. J., and Mendoza, C. A. (2004). Coupled fluid flow, heat and mass transport, and erosion in the Alberta basin: implications for the origin of the Athabasca oil sands. *Can. J. Earth Sci.* 41, 1077–1095. doi: 10.1139/e04-052

6.    Adams, J., Larter, S., Bennett, B., Huang, H., Westrich, J., and van Kruisdijk, C. (2013a). "The dynamic interplay of oil mixing, charge timing, and biodegradation in forming the Alberta oil sands: insights from geologic modeling and biogeochemistry," in *Heavy-oil and Oil-sand Petroleum Systems in Alberta and Beyond: AAPG Studies in Geology 64*, eds F. J. Hein, D. Leckie, S. Larter, and J. R. Suter (Tulsa: AAPG), 23–102.

7.    Adams, M. M., Hoarfrost, A. L., Bose, A., Joye, S. B., and Girguis, P. R. (2013b). Anaerobic oxidation of short-chain alkanes in hydrothermal sediments: potential influences on sulfur cycling and microbial diversity. *Front. Microbiol.* 4:110. doi: 10.3389/fmicb.2013.00110.

8.    Aeckersberg, F., Rainey, F. A., and Widdel, F. (1998). Growth, natural relationships, cellular fatty acids and metabolic adaptation of sulphate-reducing bacteria that utilize long-chain alkanes under anoxic conditions. *Arch. Microbiol.* 170, 361–369. doi: 10.1007/s002030050654.

9.    Agrawal, A., Vanbroekhoven, K., and Lal, B. (2010). Diversity of culturable sulfidogenic bacteria in two oil-water separation tanks in the north-eastern oil fields of India. *Anaerobe* 16, 12–18. doi: 10.1016/j.anaerobe.2009.04.005.

10.    Aitken, C. M., Jones, D. M., and Larter, S. R. (2004). Anaerobic hydrocarbon biodegradation in deep subsurface oil reservoirs. *Nature* 431, 291–294. doi: 10.1038/nature02922.

11.    Aitken, C. M., Jones, D. M., Maguire, M. J., Gray, N. D., Sherry, A., Bowler, B. F. J., et al. (2013). Evidence that crude oil alkane activation proceeds by different mechanisms under sulfate-

reducing and methanogenic conditions. *Geochim. Cosmochim. Acta* 109, 162–174. doi: 10.1016/j.gca.2013.01.031.

12. Allen, J. P., Atekwana, E. A., Atekwana, E. A., Duris, J. W., Werkema, D. D., and Rossbach, S. (2007). The microbial community structure in petroleum-contaminated sediments corresponds to geophysical signatures. *Appl. Environ. Microbiol* 73, 2860–2870. doi: 10.1128/AEM.01752-06.

13. Aller, R. C., and Rude, P. D. (1988). Complete oxidation of solid phase sulfides by manganese and bacteria in anoxic marine sediments. *Geochim. Cosmochim. Acta* 52, 751–765. doi: 10.1016/0016-7037(88)90335-3.

14. An, D., Caffrey, S. M., Soh, J., Agrawal, A., Brown, D., Budwill, K., et al. (2013). Metagenomics of hydrocarbon resource environments indicates aerobic taxa and genes to be unexpectedly common. *Environ. Sci. Technol.* 47, 10708–10717. doi: 10.1021/es4020184.

15. Atlas, R. M. (1975). Effects of temperature and crude oil composition on petroleum biodegradation. *Appl. Microbiol.* 30, 396–403.

16. Atlas, R. M., and Bartha, R. (1972a). Degradation and mineralization of petroleum in seawater: limitation by nitrogen and phosphorus. *Biotechnol. Bioeng.* 14, 309–318. doi: 10.1002/bit.260140304.

17. Atlas, R. M., and Bartha, R. (1972b). Biodegradation of petroleum in seawater at low temperatures. *Can. J. Microbiol.* 18, 1851–1855. doi: 10.1139/m72-289.

18. Bakermans, C., and Madsen, E. L. (2002). Diversity of 16S rDNA and naphthalene dioxygenase genes from coal-tar-waste-contaminated aquifer waters. *Microb. Ecol.* 44, 95–106. Available online at: http://www.jstor.org/stable/4287636

19. Ball, J. S., Whisman, M. L., and Wenger, W. J. (1951). Nitrogen content of crude, petroleums. *Ind. Eng. Chem.* 43, 2577–2581. doi: 10.1021/ie50503a047

20. Beller, H. R., and Edwards, E. A. (2000). Anaerobic toluene activation by benzylsuccinate synthase in a highly enriched methanogenic culture. *Appl. Environ. Microbiol.* 66, 5503–5505. doi: 10.1128/AEM.66.12.5503-5505.2000

21. Bennett, B., Adams, J. J., Gray, N. D., Sherry, A., Oldenburg, T. B. P., Huang, H., et al. (2013). The controls on the composition of biodegraded oils in the deep subsurface - Part 3. The impact of microorganism distribution on petroleum geochemical gradients in biodegraded petroleum reservoirs. *Org. Geochem.* 56, 94–105. doi: 10.1016/j.orggeochem.2012.12.011

22. Berdugo-Clavijo, C., and Gieg, L. M. (2014). Conversion of crude oil to methane by a microbial consortium enriched from oil reservoir production waters. *Front. Microbiol.* 5:197. doi: 10.3389/fmicb.2014.00197

23. Bjergbakke, E., Draganic, Z. D., Sehested, K., and Draganic, I. G. (1989). Radiolytic products in waters. Part II: computer simulation of some radiolytic processes in the laboratory. *Radiochimica Acta* 48, 65–71.

24. Blumenstein, I. O., Krooss, B. M., di Primio, R., Rottke, W., Müller, E., Westerlage, C., et al. (2008). Biodegradation in numerical basin modelling: a case study from the Gifhorn Trough, N-Germany. *Int. J. Earth. Sci. (Geol. Rundsch.)* 97, 1115–1129. doi: 10.1007/s00531-007-0272-1

25. Boll, M., and Heider, J. (2010). "Anaerobic degradation of hydrocarbons: mechanisms of C-H-bond activation in the absence of oxygen," in *Handbook of Hydrocarbon and Lipid Microbiology*, Volume 2 Part 4 Chapter 15, eds K. N. Timmis, T. McGenity, J. R. van der Meer, and V. de Lorenzo (Heidelberg: Springer), 1011–1024. doi: 10.1007/978-3-540-77587-4_71

26. Broecker, W. (2013). Does air capture constitute a viable backstop against a bad $CO_2$ trip? *Elementa Sci. Anthropocene*1:000009. doi: 10.12952/journal.elementa.000009

27. Callaghan, A. V. (2013). Enzymes involved in the anaerobic oxidation of n-alkanes: from methane to long-chain paraffins. *Front. Microbiol.* 4:89. doi: 10.3389/fmicb.2013.00089

28. Callaghan, A. V., Morris, B. E., Pereira, I. A., McInerney, M. J., Austin, R. N., Groves, J. T., et al. (2012). The genome sequence of *Desulfatibacillum alkenivorans* AK-01: a blueprint for anaerobic alkane oxidation. *Environ. Microbiol.* 14, 101–113. doi: 10.1111/j.1462-2920.2011.02516.x

29. Campbell, B. J., Engel, A. S., Porter, M. L., and Takai, K. (2006). The versatile ε-proteobacteria: key players in sulphidic habitats.

Nat. Rev. Microbiol. 4, 458–468. doi: 10.1038/nrmicro1414

30.   Chen, C. I., and Taylor, R. T. (1997). Thermophilic biodegradation of BTEX by two consortia of anaerobic bacteria. Appl. Microbiol. Biotechnol. 48, 121–128. doi: 10.1007/s002530051026

31.   Cheng, L., Ding, C., Li, Q., He, Q., Dai, L.-R., and Zhang, H. (2013a). DNA-SIP reveals that Syntrophaceae play an important role in methanogenic hexadecane degradation. PLoS ONE 8:e66784. doi: 10.1371/journal.pone.0066784

32.   Cheng, L., He, Q., Ding, C., Dai, L.-R., Li, Q., and Zhang, H. (2013b). Novel bacterial groups dominate in a thermophilic methanogenic hexadecane-degrading consortium. FEMS Microbiol. Ecol. 85, 568–577. doi: 10.1111/1574-6941.12141

33.   Cheng, L., Rui, J., Li, Q., Zhang, H., and Lu, L. (2013c). Enrichment and dynamics of novel syntrophs in a methanogenic hexadecane-degrading culture from a Chinese oilfield. FEMS Microbiol. Ecol. 83, 757–766. doi: 10.1111/1574-6941.12031

34.   Cheng, S., and Logan, B. E (2007). Sustainable and efficient biohydrogen production via electrohydrogenesis. Proc. Natl. Acad. Sci. U.S.A. 104, 18871–18873. doi: 10.1073/pnas.0706379104

35.   Chu, S., and Marjumbdar, A. (2012). Opportunities and challenges for a sustainable energy future. Nature 488, 294–303. doi: 10.1038/nature11475

36.   Connan, J. (1984). "Biodegradation of crude oils in reservoirs," in Advances in Petroleum Geochemistry, Vol. 1, eds J. Brooks and D. H. Welte (London: Academic), 299–335. doi: 10.1016/B978-0-12-032001-1.50011-0

37.   Creaney, S., Allan, J., Cole, K. S., Fowler, M. G., Brooks, P. W., Osadetz, K. G., et al. (1994). "Petroleum generation and migration in the Western Canada sedimentary basin," in Geological Atlas of the Western Canada Sedimentary Basin, eds G. Mossop and I. Shetson (Edmonton: Canadian Society of Petroleum Geologists and Alberta Research Council), 455–468.

38.   da Cruz, G. F., de Vasconcellos, S. P., Angolini, C. F., Dellagnezze, B. M., Garcia, I. N., de Oliveira, V. M., et al. (2011). Could petroleum biodegradation be a joint achievement of aerobic and anaerobic microorganisms in deep sea reservoirs? AMB Express 1:47 doi: 10.1186/2191-0855-1-47

39.  Dahle, H., Garshol, F., Madsen, M., and Birkeland, N.-K. (2008). Microbial community structure analysis of produced water from a high-temperature North Sea oil-field. *Antonie van Leeuwenhoek* 93, 37–49. doi: 10.1007/s10482-007-9177-z

40.  DasSarma, S., and Arora, P. (2002). *Halophiles, Encyclopedia of Life Sciences*, Vol. 8. London: Nature Publishing Group.

41.  de Oliveira, V. M., Sette, L. D., Marques-Simioni, K. C., and dos Santos Neto, E. V. (2008). Bacterial diversity characterization in petroleum samples from Brazilian reservoirs. *Braz. J. Microbiol.* 39, 445–452 doi: 10.1590/S1517-83822008000300007

42.  D'Hondt, S., Spivack, A. J., Pockalny, R., Ferdelman, T. G., Fischer, J. P., Kallmeyer, J., et al. (2009). Subseafloor sedimentary life in the South Pacific Gyre. *Proc. Natl. Acad. Sci. U.S.A.* 106, 11651–11656. doi: 10.1073/pnas.0811793106

43.  Dojka, M. A., Hugenholtz, P., Haack, S. K., and Pace, N. R. (1998). Microbial diversity in a hydrocarbon- and chlorinated-solvent-contaminated aquifer undergoing intrinsic bioremediation. *Appl. Environ. Microbiol.* 64, 3869–3877.

44.  Dolfing, J., Larter, S. R., and Head, I. M. (2008). Thermodynamic constraints on methanogenic crude oil biodegradation.*ISME J.* 2, 442–452. doi: 10.1038/ismej.2007.111

45.  Draganic, I. G. (1991). Radiolysis of water: a look at its origin and occurrence in the nature. *Radiat. Phys. Chem.* 72, 181–186. doi: 10.1016/j.radphyschem.2004.09.012

46.  Duncan, K. E., Gieg, L. M., Parisi, V. A., Tanner, R. S., Tringe, S. G., Bristow, J., et al. (2009). Biocorrosive thermophilic microbial communities in Alaskan North Slope oil facilities. *Environ. Sci. Technol.* 43, 7977–7984. doi: 10.1021/es9013932

47.  Edgcomb, V., Molyneaux, S. J., Saito, M. A., Lloyd, K., Boer, S., Wirsen, C. O., et al. (2004). Sulfide ameliorates metal toxicity for deep-sea hydrothermal vent archaea. *Appl. Environ. Microbiol.* 70, 2551–2555. doi: 10.1128/AEM.70.4.2551-2555.2004

48.  Embree, M., Nagarajan, H., Movahedi, N., Chitsaz, H., and Zengler, K. (2013). Single-cell genome and metatranscriptome sequencing reveal metabolic interactions of an alkane-degrading methanogenic community. *ISME J.* 8, 757–767. doi: 10.1038/ismej.2013.187

49.  Ettwig, K. F., Speth, D. R., Reimann, J., Wu, M. L., Jetten, M. S. M., and Keltjens, J. T. (2012). Bacterial oxygen production in the dark. *Front. Microbio.* 3:273. doi: 10.3389/fmicb.2012.00273

50.  Fagerbakke, K. M., Heldal, M., and Norland, S. (1996). Content of carbon, nitrogen, oxygen, sulfur and phosphorus in native aquatic and cultured bacteria. *Aquat. Microb. Ecol.* 10, 15–27. doi: 10.3354/ame010015

51.  Fedorovich, V., Knighton, M. C., Pagaling, E., Ward, F. B., Free, A., and Goryanin, I. (2009). Novel electrochemically active bacterium phylogenetically related to *Arcobacter butzleri*, isolated from a microbial fuel cell. *Appl. Environ. Microbiol.* 75, 7326–7334. doi: 10.1128/AEM.01345-09

52.  Fowler, S. J., Dong, X., Sensen, C. W., Suflita, J. M., and Gieg, L. M. (2012). Methanogenic toluene metabolism: community structure and intermediates. *Environ. Microbiol.* 14, 754–764 doi: 10.1111/j.1462-2920.2011.02631.x

53.  Fry, J. C., Horsfield, B., Sykes, R., Cragg, B. A., Heywood, C., Kim, G. T., et al. (2009). Prokaryotic populations and activities in an interbedded coal deposit, including a previously deeply buried section (1.6-2.3 km) above ~150 Ma basement rock. *Geomicrobiol. J.* 26, 163–178 doi: 10.1080/01490450902724832

54.  Fustic, M., Bennett, B., Adams, J., Huang, H., MacFarlane, B., Leckie, D. A., et al. (2011). Bitumen and heavy oil geochemistry: a tool for distinguishing barriers from baffles in oil sands reservoirs. *Bull. Can. Petroleum Geol.* 59, 295–316. doi: 10.2113/gscpgbull.59.4.295

55.  Gevertz, D., Telang, A. J., Voordouw, G., and Jenneman, G. E. (2000). Isolation and characterization of strains CVO and FWKOB, two novel nitrate-reducing, sulfide-oxidizing bacteria isolated from oil field brine. *Appl. Environ. Microbiol.* 66, 2491–2501. doi: 10.1128/AEM.66.6.2491-2501.2000

56.  Gieg, L. M., Davidova, I. A., Duncan, K. E., and Suflita, J. M. (2010). Methanogenesis, sulfate reduction and crude oil biodegradation in hot Alaskan oilfields. *Environ. Microbiol.* 12, 3074–3086. doi: 10.1111/j.1462-2920.2010.02282.x

57.  Gieg, L. M., Duncan, K. E., and Suflita, J. M. (2008). Bioenergy production via microbial conversion of residual oil to natural

gas. *Appl. Environ. Microbiol.* 74, 3022–3029. doi: 10.1128/AEM.00119-08

58.  Grabowski, A., Nercessian, O., Fayolle, F., Blanchet, D., and Jeanthon, C. (2005). Microbial diversity in production waters of a low-temperature biodegraded oil reservoir. *FEMS Microbiol. Ecol.* 54, 427–443. doi: 10.1016/j.femsec.2005.05.007

59.  Grassia, G. S., McLean, K. M., Glenat, P., Bauld, J., and Sheehy, A. J. (1996). A systematic survey for thermophilic fermentative bacteria and archaea in high-temperature petroleum reservoirs. *FEMS Microbiol. Ecol.* 21, 47–58. doi: 10.1111/j.1574-6941.1996.tb00332.x

60.  Graves, C., Ebbesen, S. D., Mogensen, M., and Lackner, K. S. (2011). Sustainable hydrocarbon fuels by recycling $CO_2$ and $H_2O$ with renewable or nuclear energy. *Renew. Sustain. Energy Rev.* 15, 1–23. doi: 10.1016/j.rser.2010.07.014

61.  Gray, N. D., Sherry, A., Grant, R. J., Rowan, A. K., Hubert, C. R. J., Callbeck, C. M., et al. (2011). The quantitative significance of Syntrophaceae and syntrophic partnerships in methanogenic degradation of crude oil alkanes. *Environ. Microbiol.* 13, 2957–2975. doi: 10.1111/j.1462-2920.2011.02570.x

62.  Gray, N. D., Sherry, A., Hubert, C., Dolfing, J., and Head, I. M. (2010). Methanogenic degradation of petroleum hydrocarbons in subsurface environments: remediation, heavy oil formation, and energy recovery. *Adv. Appl. Microbiol.* 72, 137–161. doi: 10.1016/S0065-2164(10)72005-0

63.  Gray, N. D., Sherry, A., Larter, S. R., Erdmann, M., Leyris, J., Liengen, T., et al. (2009). Biogenic methane production in formation waters from a large gas field in the North Sea. *Extremophiles* 13, 511–519. doi: 10.1007/s00792-009-0237-3

64.  Grossi, V., Cravo-Laureau, C., Guyoneaud, R., Ranchou-Peyruse, A., and Hirschler-Réa, A. (2008). Metabolism of n-alkanes and n-alkenes by anaerobic bacteria: a summary. *Org. Geochem.* 39, 1197–1203. doi: 10.1016/j.orggeochem.2008.02.010

65.  Gunderson, K., Heldal, M., Norland, S., Purdie, D. A., and Knap, A. H. (2002). Elemental C, N, and P cell content of individual bacteria collected at the Bermuda Atlantic Time-Series Study (BATS) Site. *Limnol. Oceoanogr.* 47, 1525–1530. doi: 10.4319/lo.2002.47.5.1525

66. Guo, L., Li, X.-M., Bo, X., Yang, Q., Zeng, G.-M., Liao, D.-X., et al. (2008). Impacts of sterilization, microwave and ultrasonication pretreatment on hydrogen producing using waste sludge. *Bioresour. Technol.* 99, 3651–3658. doi: 10.1016/j.biortech.2007.07.026

67. Haeseler, F., Behar, F., Garnier, D., and Chenet, P.-Y. (2010). First stoichiometric model of oil biodegradation in natural petroleum systems: Part I - the BioClass 0D approach. *Org. Geochem.* 41, 1156–1170. doi: 10.1016/j.orggeochem.2010.05.019

68. Hallmann, C., Schwark, L., and Grice, K. (2008). Community dynamics of anaerobic bacteria in deep petroleum reservoirs. *Nat. Geosci.* 1, 588–591. doi: 10.1038/ngeo260

69. Head, I. M., Jones, D. M., and Larter, S. R. (2003). Biological activity in the deep subsurface and the origin of heavy oil.*Nature* 426, 344–352. doi: 10.1038/nature02134

70. Head, I. M., Jones, D. M., and Röling, W. F. M. (2006). Marine microorganisms make a meal of oil. *Nat. Rev. Microbiol.* 4, 173–182. doi: 10.1038/nrmicro1348

71. Head, I. M., Larter, S. R., Gray, N. D., Sherry, A., Adams, J. J., Aitken, C. M., et al. (2010). "Hydrocarbon degradation in petroleum reservoirs," in *Handbook of Hydrocarbon and Lipid Microbiology,* Vol. 4, Part 6, Chapter 54, eds J. R. van derMeer and V. de Lorenzo; Editor-in-chief: K. N. Timmis and T. J. McGenity (Heidelberg: Springer), 3097–3109.

72. Hein, F. J., Leckie, D., Larter, S., and Suter, J. R. (2013). "Heavy oil and bitumen petroleum systems in Alberta and beyond: the future is nonconventional and the future is now," in *Heavy-oil and Oil-sand Petroleum Systems in Alberta and Beyond: AAPG Studies in Geology 64*, eds F. J. Hein, D. Leckie, S. Larter, and J. R. Suter (Tulsa: AAPG), 1–22.

73. Herrmann, S., Kleinsteuber, S., Chatzinotas, A., Kuppardt, S., Lueders, T., Richnow, H. H., et al. (2010). Functional characterization of an anaerobic benzene-degrading enrichment culture by DNA stable isotope probing. *Environ. Microbiol.* 12, 401–411. doi: 10.1111/j.1462-2920.2009.02077.x

74. Holmes, D. E., Risso, C., Smith, J. A., and Lovley, D. R. (2011). Anaerobic oxidation of benzene by the hyperthermophilic

archaeon *Ferroglobus placidus*. *Appl. Environ. Microbiol.* 77, 5926–5933. doi: 10.1128/AEM.05452-11

75. Horstad, I., and Larter, S. R. (1997). Petroleum migration, alteration, and remigration within Troll field, Norwegian North Sea. *Am. Assoc. Pet. Geol. Bull.* 81, 222–248.

76. Huang, H., Bowler, B. F. J., Zhang, Z., Oldenburg, T. B. P., and Larter, S. R. (2003). Influence of biodegradation on carbazole and benzocarbazole distributions in oil columns from the Liaohe basin, N.E. China. *Org. Geochem.* 34, 951–969. doi: 10.1016/S0146-6380(03)00033-0

77. Hubert, C. (2010). "Microbial ecology of oil reservoir souring control by nitrate injection," in *Handbook of Hydrocarbon and Lipid Microbiology*, ed K. N. Timmis (Berlin; Heidelberg: Springer), 2753–2766. doi: 10.1007/978-3-540-77587-4_204

78. Hubert, C. R. J., Oldenburg, T. B. P., Fustic, M., Gray, N. D., Larter, S. R., Penn, K., et al. (2012). Massive dominance of*Epsilonproteobacteria* in formation waters from a Canadian oil sands reservoir containing severely biodegraded oil.*Environ. Microbiol.* 14, 387–404. doi: 10.1111/j.1462-2920.2011.02521.x

79. Johnson, H. A., Pelletier, D. A., and Spormann, A. M. (2001). Isolation and characterization of anaerobic ethylbenzene dehydrogenase, a novel Mo-Fe-S enzyme. *J. Bacteriol.* 183, 4536–4542. doi: 10.1128/JB.183.15.4536-4542.2001

80. Jones, D. M., Head, I. M., Gray, N. D., Adams, J. J., Rowan, A. K., Aitken, C. M., et al. (2008). Crude oil biodegradation via methanogenesis in subsurface petroleum reservoirs. *Nature* 451, 176–180. doi: 10.1038/nature06484

81. Jorgensen, B. B. (1990). A thiosulfate shunt in the sulfur cycle of marine sediments. *Science* 249, 152–154. doi: 10.1126/science.249.4965.152

82. Kasai, Y., Takahata, Y., Hoaki, T., and Watanabe, K. (2005). Physiological and molecular characterization of a microbial community established in unsaturated, petroleum-contaminated soil. *Environ. Microbiol.* 7, 806–818. doi: 10.1111/j.1462-2920.2005.00754.x

83. Kashefi, K., and Lovley, D. R. (2003). Extending the upper temperature limit for life. *Science* 301, 934. doi: 10.1126/science.1086823

84. Khavari-Khorasani, G., Michelsen, J. K., and Dolson, J. C. (1998). The factors controlling the abundance and migration of heavy vs. light oils, as constrained by data from the Gulf of Suez. Part II. The significance of reservoir mass transport processes. *Org. Geochem.* 29, 283–300 doi: 10.1016/S0146-6380(98)00151-X

85. Khelifi, N., Amin Ali, O., Roche, P., Grossi, V., Brochier-Armanet, C., Valette, O., et al. (2014). Anaerobic oxidation of long-chain n-alkanes by the hyperthermophilic sulfate-reducing archaeon, *Archaeoglobus fulgidus. ISME J.* 8, 2153–2166. doi: 10.1038/ismej.2014.58

86. Khelifi, N., Grossi, V., Hamdi, M., Dolla, A., Tholozan, J. L., Ollivier, B., et al. (2010). Anaerobic oxidation of fatty acids and alkenes by the hyperthermophilic sulfate-reducing archaeon *Archaeoglobus fulgidus. Appl. Environ. Microbiol.* 76, 3057–3060. doi: 10.1128/AEM.02810-09

87. Kim, H., Choo, Y.-J., Song, J., Lee, J.-S., Lee, K. C., and Cho, J.-C. (2007). *Marinobacterium litorale* sp. nov. in the order*Oceanospirillales. Int. J. Syst. Evol. Microbiol.* 57, 1659–1662. doi: 10.1099/ijs.0.64892-0

88. Kniemeyer, O., Musat, F., Sievert, S. M., Knittel, K., Wilkes, H., Blumenberg, M., et al. (2007). Anaerobic oxidation of short-chain hydrocarbons by marine sulphate-reducing bacteria. *Nature* 449, 898–901. doi: 10.1038/nature06200

89. Kobayashi, H., Endo, K., Sakata, S., Mayumi, D., Kawaguchi, H., Ikarashi, M., et al. (2012). Phylogenetic diversity of microbial communities associated with the crude-oil, large-insoluble-particle and formation-water components of the reservoir fluid from a non-flooded high-temperature petroleum reservoir. *J. Biosci. Bioeng.* 113, 204–210. doi: 10.1016/j.jbiosc.2011.09.015

90. Kodama, Y., Ha, L. T., and Watanabe, K. (2007). *Sulfurospirillum cavolei* sp. nov., a facultatively anaerobic sulfur-reducing bacterium isolated from an underground crude oil storage cavity. *Int. J. Syst. Evol. Microbiol.* 57, 827–831. doi: 10.1099/ijs.0.64823-0

91. Kodama, Y., and Watanabe, K. (2003). Isolation and characterization of a sulfur-oxidizing chemolithotroph growing on crude oil under anaerobic conditions. *Appl. Environ. Microbiol.* 69, 107–112. doi: 10.1128/AEM.69.1.107-112.2003

92.  Kodama, Y., and Watanabe, K. (2004). *Sulfuricurvum kujiense* gen. nov., sp. nov., a facultatively anaerobic, chemolithoautotrophic, sulfur-oxidizing bacterium isolated from an underground crude-oil storage cavity. *Int. J. Syst. Evol. Microbiol.* 54, 2297–2300. doi: 10.1099/ijs.0.63243-0

93.  Kryachko, Y., Dong, X., Sensen, C. W., and Voordouw, G. (2012). Compositions of microbial communities associated with oil and water in a mesothermic oil field. *Antonie van Leeuwenhoek* 101, 493–506. doi: 10.1007/s10482-011-9658-y

94.  Lackner, K. S., Brennan, S., Matter, J. M., Park, A.-H. A., Wright, A., and van der Zwaan, B. (2012). The urgency of the development of $CO_2$ capture from ambient air. *Proc. Natl. Acad. Sci. U.S.A.* 109, 13156–13162. doi: 10.1073/pnas.1108765109

95.  Larter, S., Adams, J., Gates, I. D., Bennett, B., and Huang, H. (2008). The origin, prediction and impact of oil viscosity heterogeneity on the production characteristics of tar sand and heavy oil reservoirs. *J. Can. Petroleum Technol.* 47, 52–61. doi: 10.2118/08-01-52

96.  Larter, S., and Head, I. (2014). Oil sands and heavy oil: origin and exploitation. *Elements* 10, 277–283. doi: 10.2113/gselements.10.4.277

97.  Larter, S., Huang, H., Adams, J., Bennett, B., Jokanola, O., Oldenburg, T., et al. (2006). The controls on the composition of biodegraded oils in the deep subsurface: Part II - Geological controls on subsurface biodegradation fluxes and constraints on reservoir-fluid property prediction. *Am. Assoc. Pet. Geol. Bull.* 90, 921–938. doi: 10.1306/01270605130

98.  Larter, S. R., Head, I. M., Huang, H., Bennett, B., Jones, M., Aplin, A. C., et al. (2005). "Biodegradation, gas destruction and methane generation in deep subsurface petroleum reservoirs," in *Petroleum Geology: North-West Europe and Global Perspectives*, eds A. G. Doré and B. A. Vining (London: Proceedings Geology Conferences Ltd. Published by the Geological Society), 633–639.

99.  Larter, S. R., Head, I. M., Jones, D. M., Erdmann, M., and Wilhelms, A. (2012). *Process for Stimulating Production of Hydrogen from Petroleum in Subterranean Formations*. Gatineau: Canadian IP Office.

100. Larter, S., Wilhelms, A., Head, I., Koopmans, M., Aplin, A., Di Primio, R., et al. (2003). The controls on the composition of biodegraded oils in the deep subsurface - part 1: biodegradation rates in petroleum reservoirs. *Org. Geochem.* 34, 601–613. doi: 10.1016/S0146-6380(02)00240-1

101. Li, D., Midgley, D. J., Ross, J. P., Oytam, Y., Abell, G. C., Volk, H., et al. (2012). Microbial biodiversity in a Malaysian oil field and a systematic comparison with oil reservoirs worldwide. *Arch. Microbiol.* 194, 513-523. doi: 10.1007/s00203-012-0788-z

102. Lin, L.-H., Slater, G. F., Sherwood-Lollar, B., Lacrampe-Couloume, G., and Onstott, T. C. (2005). The yield and isotopic composition of radiolytic $H_2$, a potential energy source for the deep subsurface biosphere. *Geochim. Cosmochim. Acta* 69, 893–903. doi: 10.1016/j.gca.2004.07.032

103. Lloyd, K., Edgcomb, V. P., Molyneaux, S. M., Boer, S., Wirsen, C. O., Atkins, M., et al. (2005). Effect of dissolved sulfide, pH, and temperature on the growth and survival of marine hyperthermophilic archaea. *Appl. Environ. Microbiol.* 71, 6383–6387. doi: 10.1128/AEM.71.10.6383-6387.2005

104. Luijten, M. L. G. C., de Weert, J., Smidt, H., Boschker, H. T. S., de Vos, W. M., Schraa, G., et al. (2003). Description of *Sulfurospirillum halorespirans* sp. nov., an anaerobic, tetrachloroethene-respiring bacterium, and transfer of *Dehalospirillum multivorans* to the genus *Sulfurospirillum* as *Sulfurospirillum multivorans* comb. nov. *Int. J. Syst. Evol. Microbiol.* 53, 787–793. doi: 10.1099/ijs.0.02417-0

105. Magot, M. (2005). "Indigenous microbial communities in oil fields," in *Petroleum Microbiology*, eds B. Ollivier and M. Magot (Washington, DC: ASM Press), 21–33.

106. Magot, M., Ollivier, B., and Patel, B. K. C. (2000). Microbiology of petroleum reservoirs. *Antonie van Leeuwenhoek* 77, 103–116. doi: 10.1023/A:1002434330514

107. Manning, D. A. C., and Hutcheon, I. (2004). Distribution and mineralogical controls on ammonium in deep groundwaters. *Appl. Geochem.* 19, 1495–1503. doi: 10.1016/j.apgeochem.2004.01.019

108. Mardanov, A. V., Ravin, N. V., Svetlitchnyi, V. A., Beletsky, A. V., Miroshnichenko, M. L., Bonch-Osmolovskaya, E. A., et al. (2009).

Metabolic versatility and indigenous origin of the archaeon *Thermococcus sibiricus*, isolated from a Siberian oil reservoir, as revealed by genome analysis. *Appl. Environ. Microbiol.* 75, 4580–4588. doi: 10.1128/AEM.00718-09

109. Mayumi, D., Dolfing, J., Sakata, S., Maeda, H., Miyagawa, Y., Ikarashi, M., et al. (2013). Carbon dioxide concentration dictates alternative methanogenic pathways in oil reservoirs. *Nat. Commun.* 4, 1998. doi: 10.1038/ncomms2998

110. Mbadinga, S. M., Li, K.-P., Zhou, L., Wang, L.-Y., Yang, S.-Z., Liu, J.-F., et al. (2012). Analysis of alkane-dependent methanogenic community derived from production water of a high-temperature petroleum reservoir. *Appl. Microbiol. Biotechnol.* 96, 531–542. doi: 10.1007/s00253-011-3828-8

111. Meslé, M., Dromart, G., and Oger, P. (2013). Microbial methanogenesis in subsurface oil and coal. *Res. Microbiol.* 164, 959–972. doi: 10.1016/j.resmic.2013.07.004

112. Milkov, A. V. (2011). Worldwide distribution and significance of secondary microbial methane formed during petroleum biodegradation in conventional reservoirs. *Org. Geochem.* 42, 184–207. doi: 10.1016/j.orggeochem.2010.12.003

113. Moldowan, J. M., and McCaffrey, M. A. (1995). A novel microbial hydrocarbon degradation pathway revealed by hopane demethylation in a petroleum reservoir. *Geochim. Cosmochim. Acta* 59, 1891–1894. doi: 10.1016/0016-7037(95)00072-8

114. Morono, Y., Terada, T., Nishizawa, M., Ito, M., Hillion, F., Takahata, N., et al. (2011). Carbon and nitrogen assimilation in deep subseafloor microbial cells. *Proc. Natl. Acad. Sci. U.S.A.* 108, 18295–18300. doi: 10.1073/pnas.1107763108

115. Oldenburg, T. B. P., Larter, S. R., Adams, J. J., Hubert, C., Rowan, A., Brown, A., et al. (2009). Methods for recovery of microorganisms and intact polar lipids (IPLs) from oil-water mixtures-lab experiments and natural well-head fluids. *Anal. Chem.* 81, 4130–4136. doi: 10.1021/ac8025515

116. Oldenburg, T. B. P., Larter, S. R., and Huang, H. (2006). Nutrient supply during subsurface oil biodegradation - availability of petroleum nitrogen as a nutrient source for subsurface microbial activity. *Energy Fuels* 20, 2079–2082. doi: 10.1021/ef060148p

117. Olivier, B., Fardeau, M.-L., Cayol, J.-L., Magot, M., Patel, B. K. C., Prensier, G., et al. (1998). *Methanocalculus halotolerans* gen. nov., sp. nov., isolated from an oil-producing well. *Int. J. Syst. Evol. Microbiol.* 48, 821–828.

118. Oren, A. (1999). Bioenergetic aspects of halophilism. *Microbiol. Mol. Biol. Rev.* 63, 334–348.

119. Oren, A. (2010). Thermodynamic limits to microbial life at high salt concentrations. *Environ. Microbiol.* 13, 1908–1923. doi: 10.1111/j.1462-2920.2010.02365.x

120. Orphan, V. J., Boles, J. R., Goffredi, S. K., and Delong, E. F. (2003). Geochemical influence on community structure and microbial processes in high temperature oil reservoirs. *Geomicrobiol. J.* 20, 295–311. doi: 10.1080/01490450303898

121. Orphan, V. J., Taylor, L. T., Hafenbradl, D., and Delong, E. F. (2000). Culture-dependent and culture-independent characterization of microbial assemblages associated with high-temperature petroleum reservoirs. *Appl. Environ. Microbiol.* 66, 700–711. doi: 10.1128/AEM.66.2.700-711.2000

122. Palmer, E. S. (1993). "Effect of biodegradation and water washing on crude oil composition," in *Organic Geochemistry Principles and Applications*, eds M. Engel and S. A. Macko (New York, NY: Plenum Press), 511–534. doi: 10.1007/978-1-4615-2890-6_23

123. Parkes, R. J., Cragg, B. A., and Wellsbury, P. (2000). Recent studies on bacterial populations and processes in subseafloor sediments: a review. *Hydrogeol. J.* 8, 11–28. doi: 10.1007/PL00010971

124. Peters, K. E., and Moldowan, J. M. (1993). *The Biomarker Guide: Interpreting Molecular Fossils in Petroleum and Ancient Sediments*. New Jersey, NJ: Prentice Hall, Englewood Cliffs.

125. Peters, K. E., Walters, C. C., and Moldowan, J. M. (2005). *The Biomarker Guide*. 2nd Edn. Cambridge: Cambridge University Press.

126. Rabaey, K., Boon, N., Höfte, M., and Verstraete, W. (2005). Microbial phenazine production enhances electron transfer in biofuel cells. *Environ. Sci. Technol.* 39, 3401–3408. doi: 10.1021/es048563o

127. Ramos-Padrón, E., Bordenave, S., Lin, S., Bhaskar, I. M., Dong, X., Sensen, C. W., et al. (2011). Carbon and sulfur cycling by

microbial communities in a gypsum-treated oil sands tailings pond. *Environ. Sci. Technol.* 45, 439–446. doi: 10.1021/es1028487

128. Redfield, A. C. (1934). "On the proportions of organic derivations in sea water and their relation to the composition of plankton," in *James Johnstone Memorial Volume*, ed R. J. Daniel (Liverpool: University Press of Liverpool), 177–192.

129. Redfield, A. C. (1958). The biological control of chemical factors in the environment. *Am. Sci.* 46, 205–221.

130. Richter, P. P., Ceasser, P. D., Meisel, S. A., and Offenhauser, R. D. (1952). Distribution of nitrogen, according to basicity.*Ind. Eng. Chem.* 44, 2601–2605. doi: 10.1021/ie50515a037

131. Rodriguez-Martinez, M. F., Kelessidou, N., Law, Z., Gardiner, J., and Stephens, G. (2008). Effect of solvents on obligately anaerobic bacteria. *Anaerobe* 14, 55–60. doi: 10.1016/j.anaerobe.2007.09.006

132. Rogers, J. R., Bennett, P. C., and Choi, W. J. (1998). Feldspars as a source of nutrients for microorganisms. *Am. Mineral.* 83, 1532–1540.

133. Röling, W. F. M., Larter, S. R., and Head, I. M. (2003). The microbiology of hydrocarbon degradation in subsurface petroleum reservoirs: perspectives and prospects. *Res. Microbiol.* 154, 321–328. doi: 10.1016/S0923-2508(03)00086-X

134. Roussel, E. G., Cambon-Bonavita, M. A., Querellou, J., Cragg, B. A., Webster, G., Prieur, D., et al. (2008). Extending the sub-sea-floor biosphere. *Science* 320, 1046. doi: 10.1126/science.1154545

135. Rueter, P., Rabus, R., Wilkes, H., Aeckersberg, F., Rainey, F. A., Jannasch, H. W., et al. (1994). Anaerobic oxidation of hydrocarbons in crude oil by new types of sulfate reducing bacteria. *Nature* 372, 455–458. doi: 10.1038/372455a0

136. Sereda, N., and James, B. R. (2014). "A case study in the application of bitumen geochemistry for reservoir characterization in SAGD development," in *SPE Heavy Oil Conference-Canada* (Houston: SPE). doi: 10.2118/170109-MS

137. Sette, L. D., Simioni, K. C. M., Vasconcellos, S. P., Dussan, L. J., Neto, E. V. S., and Oliveira, V. M. (2007). Analysis of the

composition of bacterial communities in oil reservoirs from a southern offshore Brazilian basin. *Antonie van Leeuwenhoek* 91, 253–266. doi: 10.1007/s10482-006-9115-5

138. Sherry, A., Grant, R. J., Aitken, C. M., Jones, D. M., Head, I. M., and Gray, N. D. (2014). Volatile hydrocarbons inhibit methanogenic crude oil degradation. *Front. Microbiol.* 5:131. doi: 10.3389/fmicb.2014.00131

139. Sherry, A., Gray, N. D., Ditchfield, A. K., Aitken, C. M., Jones, D. M., Röling, W. F. M., et al. (2012). Anaerobic biodegradation of crude oil under sulfate-reducing conditions leads to only modest enrichment of recognized sulfate-reducing taxa. *Int. Biodeterior. Biodegradation* 81, 105–113. doi: 10.1016/j.ibiod.2012.04.009

140. Shimizu, S., Akiyama, M., Naganuma, T., Fujioka, M., Nako, M., and Ishijima, Y. (2007). Molecular characterization of microbial communities in deep coal seam groundwater of northern Japan. *Geobiology* 5, 423–433. doi: 10.1111/j.1472-4669.2007.00123.x

141. Siddique, T., Penner, T., Semple, K., and Foght, J. M. (2011). Anaerobic biodegradation of longer-chain n-alkanes coupled to methane production in oil sands tailings. *Environ. Sci. Technol.* 45, 5892–5899 doi: 10.1021/es200649t

142. So, C. M., Phelps, C. D., and Young, L. Y. (2003). Anaerobic transformation of alkanes to fatty acids by a sulphate-reducing bacterium, strain Hxd3. *Appl. Environ. Microbiol.* 69, 3892–3900. doi: 10.1128/AEM.69.7.3892-3900.2003

143. Stainford, J. G. (2004). "New insights into reservoir filling and mixing processes," in *Understanding Petroleum Reservoirs: Towards an Integrated Reservoir Engineering and Geochemical Approach*, eds J. M. Cubitt, W. A. England, S. Larter (London: Geological Society, Special Publications), 237, 115–132.

144. Stetter, K. O., Huber, R., Blochl, E., Kurr, M., Eden, R. D., Fielder, M., et al. (1993). Hyperthermophilic archaea are thriving in deep North Sea and Alaskan oil reservoirs. *Nature* 365, 743–745. doi: 10.1038/365743a0

145. Stolaroff, J. K., Keith, D. W., and Lowry, G. V. (2008). Carbon dioxide capture from atmospheric air using sodium hydroxide spray. *Environ. Sci. Technol.* 42, 2728–2735 doi: 10.1021/es702607w

146. Strausz, O. P., and Lown, E. M. (2003). *The Chemistry of Alberta Oil Sands Bitumens and Heavy Oils*. Calgary, AB: The Alberta Energy Research Institute.

147. Strous, M. (2014). *Innovative Strategies to Reduce the Cost of Carbon Capture Biotechnology*. Available online at:http://www.ucalgary.ca/utoday/issue/2014-05-09/researchers-believe-cheaper-algae-biotechnology-can-help-environment

148. Sunde, E., and Torsvik, T. (2005). "Microbial control of hydrogen sulfide production in oil reservoirs," in *Petroleum Microbiology*, eds B. Ollivier and M. Magot (Washington DC: ASM Press), 201–213. doi: 10.1128/9781555817589.ch10

149. Takai, K., Nakamura, K., Tomohiro, T., Tsunogai, U., Miyazaki, M., Miyazaki, J., et al. (2008). Cell proliferation at 122°C and isotopically heavy $CH_4$ production by a hyperthermophilic methanogen under high-pressure cultivation. *Proc. Natl. Acad. Sci. U.S.A.* 105, 10949–10954. doi: 10.1073/pnas.0712334105

150. Tan, B., Nesbo, C., and Foght, J. (2014). Re-analysis of omics data indicates *Smithella* may degrade alkanes by addition to fumarate under methanogenic conditions. *ISME J.* doi: 10.1038/ismej.2014.87. [Epub ahead of print].

151. Taylor, P. N., Bennett, B., Jones, D. M., and Larter, S. R. (2001). The effect of biodegradation and water washing on the occurrence of alkylphenols in crude oils. *Org. Geochem.* 32, 341–358. doi: 10.1016/S0146-6380(00)00176-5

152. Thamdrup, B., Rossello-Mora, R., and Amann, R. (2000). Microbial manganese and sulfate reduction in Black Sea shelf sediments. *Appl. Environ. Microbiol.* 66, 2888–2897. doi: 10.1128/AEM.66.7.2888-2897.2000

153. Townsend, G. T., Prince, R. C., and Suflita, J. M. (2003). Anaerobic oxidation of crude oil hydrocarbons by the resident microorganisms of a contaminated anoxic aquifer. *Environ. Sci. Technol.* 37, 5213–5218. doi: 10.1021/es0264495

154. Tozer, R. S. J., Choi, A. P., Pietras, J. T., and Tanasichuk, D. J. (2014). Athabasca oil sands: megatrap restoration and charge timing. *Am. Assoc. Pet. Geol. Bull.* 98, 429–447. doi: 10.1306/08071313039

155. van der Kraan, G., Bruining, J., Lomans, B. P., van Loosdrecht, M. C. M., and Muyzer, G. (2010). Microbial diversity of an oil-

water processing site and its associated oil field: the possible role of microorganisms as information carriers from oil-associated environments. *FEMS Microbiol. Ecol.* 71, 428–443. doi: 10.1111/j.1574-6941.2009.00813.x

156. Vrede, K., Heldal, M., Norland, S., and Bratbak, G. (2002). Elemental composition (C, N, P) and cell volume of exponentially growing and nutrient-limited bacterioplankton. *Appl. Environ. Microbiol.* 68, 2965–2971. doi: 10.1128/AEM.68.6.2965-2971.2002

157. Walker, J. D., Petrakis, L., and Colwell, R. R. (1976). Comparison of the biodegradability of crude and fuel oils. *Can. J. Microbiol.* 22, 598–602. doi: 10.1139/m76-089

158. Wang, L.-Y., Gao, C.-X., Mbadinga, S. M., Zhou, L., Liu, J.-F., Gu, J.-D., et al. (2011). Characterization of an alkane-degrading methanogenic enrichment culture from production water of an oil reservoir after 274 days of incubation. *Int. Biodeterior. Biodegrad.* 65, 444–450 doi: 10.1016/j.ibiod.2010.12.010

159. Wang, Z., Fingas, M., Blenkinsopp, S., Sergy, G., Landriault, M., Sigouin, L., et al. (1998). Comparison of oil composition changes due to biodegradation and physical weathering in different oils. *J. Chromatogr. A* 809, 89–107. doi: 10.1016/S0021-9673(98)00166-6

160. Warren, E., Bekins, B., Godsy, E., and Smith, V. (2003). Inhibition of acetoclastic methanogenesis in crude oil- and creosote-contaminated groundwater. *Bioremediat. J.* 7, 139–149. doi: 10.1080/713607984

161. Warren, E., Bekins, B., Godsy, E., and Smith, V. (2004). Inhibition of acetoclastic methanogenesis in crude oil- and creosote-contaminated groundwater. *Bioremediat. J.* 8, 1–11. doi: 10.1080/10889860490465840

162. Webster, G., Rinna, J., Roussel, E. G., Fry, J. C., Weightman, A. J., and Parkes, R. J. (2010). Prokaryotic functional diversity in different biogeochemical depth zones in tidal sediments of the Severn estuary, UK, revealed by stable-isotope probing. *FEMS Microbiol. Ecol.* 72, 179–197. doi: 10.1111/j.1574-6941.2010.00848.x

163. Westerholm, M., Dolfing, J., Sherry, A., Gray, N. D., Head, I. M., and Schnürer, A. (2011). Quantification of syntrophic acetate-oxidizing microbial communities in biogas processes.

*Environ. Microbiol. Rep.* 3, 500–505. doi: 10.1111/j.1758-2229.2011.00249.x

164. Westlake, D. W. S., Jobson, A., Phillippe, R., and Cook, F. D. (1974). Biodegradability and crude oil composition. *Can. J. Microbiol.* 20, 915–928. doi: 10.1139/m74-141

165. Whittle, K. J., Hardy, P. R., Mackie, A. S., McGill, D. S., Crisp, D. J., Baker, J. M., et al. (1982). A quantitative assessment of the sources and fate of petroleum compounds in the marine environment. *Philos. Trans. R. Soc. B Biol. Sci.* 297, 193–218. doi: 10.1098/rstb.1982.0038

166. Wilhelms, A., Larter, S. R., Head, I., Farrimond, P., di-Primio, R., and Zwach, C. (2001). Biodegradation of oil in uplifted basins prevented by deep-burial sterilisation. *Nature* 411, 1034–1037. doi: 10.1038/35082535

167. Xie, X., Ye, M., Hsu, P.-C., Liu, N., Criddle, C. S., and Cui, Y. (2013). Microbial battery for efficient energy recovery. *Proc. Natl. Acad. Sci. U.S.A.* 110, 15925–15930 doi: 10.1073/pnas.1307327110

168. Yuehui, S., Beiwen, Z., Fan, Z., Zhengliang, W., Fuchang, S., Lingge, Z., et al. (2008). "Analysis on the microbial diversity of Qinghai high salt content oil reservoirs," in *The 2nd International Conference on Bioinformatics and Biomedical Engineering* (Piscataway: IEEE), 780–783.

169. Zeman, F. S., and Keith, D. W. (2008). Carbon neutral hydrocarbons. *Philos. Trans. R. Soc. Lond. A* 366, 3901–3918 doi: 10.1098/rsta.2008.0143

170. Zengler, K., Richnow, H. H., Rossello-Mora, R., Michaelis, W., and Widdel, F. (1999). Methane formation from long-chain alkanes by anaerobic microorganisms. *Nature* 401, 266–269. doi: 10.1038/45777

171. Zhang, F., She, Y. H., Chai, L. J., Banat, I. M., Zhang, X. T., Shu, F. C., et al. (2012). Microbial diversity in long-term water-flooded oil reservoirs with different *in situ* temperatures in China. *Sci. Rep.* 2:760. doi: 10.1038/srep00760

172. Zhilina, T. N., Zavarzina, D. G., Kevbrin, V. V., and Kolganova, T. V. (2013). *Methanocalculus natronophilus* sp. nov., a new alkaliphilic hydrogenotrophic methanogenic archaeon from a soda lake, and proposal of the new family Methanocalculaceae. *Microbiology* 82, 698–706. doi: 10.1134/S0026261713060131

173. Zhou, L., Li, K.-P., Mbadinga, S. M., Yang, S.-Z., Gu, J.-D., and Mu, B.-Z. (2012). Analyses of n-alkanes degrading community dynamics of a high-temperature methanogenic consortium enriched from production water of a petroleum reservoir by a combination of molecular techniques. *Ecotoxicology* 21, 1680–1691. doi: 10.1007/s10646-012-0949-5.

# Compositional Simulation on the Flow of Polymeric Solution Alternating CO2 through Heavy Oil Reservoir

Moon Sik Jeong, Jinhyung Cho, Jinsuk Choi,
Ji Ho Lee, and Kun Sang Lee

Department of Natural Resources and Environmental Engineering, Hanyang University, 222 Wangsimni-ro, Seongdong-gu, Seoul 133-791, Republic of Korea

## ABSTRACT

Water-alternating-gas (WAG) method provides superior mobility control of $CO_2$ and improves sweep efficiency. However, WAG process has some problems in highly viscous oil reservoir such as gravity overriding and poor mobility ratio. To examine the applicability of carbon dioxide to recover viscous oil from highly heterogeneous reservoirs, this study suggests polymer-alternating-gas (PAG) process. The process involves a combination of polymer flooding and $CO_2$ injection. In this numerical model, high viscosity of oil and high heterogeneity of reservoir are the main challenges. To confirm the effectiveness of PAG process in the model, four processes (waterflooding, continuous $CO_2$ injection, WAG process, and PAG process) are implemented and recovery factor, WOR, and GOR are compared. Simulation results show that PAG method would increase oil recovery over 45% compared with WAG process. The WAG ratio of 2 is found to be the optimum value for maximum oil recovery. The additional oil recovery of 3% through the 2 WAG ratio is achieved over the base case of 1:1 PAG ratio and 180 days cycle period.

# INTRODUCTION

Recently, interest in $CO_2$ flooding has grown as a method of enhanced heavy oil recovery. Injected $CO_2$ can extract the heavy oil components by oil swelling and viscosity reduction. However, the mobility ratio of $CO_2$ is unfavorable to recover heavy oils. It causes viscosity fingering and gravity override through heterogeneous reservoirs. These phenomena make an early breakthrough of injected $CO_2$ and reduce oil recovery. The problems led by poor viscosity ratio are more severe in heavy oils than light oils. Although the $CO_2$ flooding has been applied and its success has been reported in many heavy oil cases [1–7], there still remain the aforementioned problems that need to be solved in order to implement the $CO_2$ injection in the heavy oil reservoirs.

Mobility control in $CO_2$ flooding is very important to solve the low recovery efficiency problem. $CO_2$ injection method can achieve higher microscopic displacement efficiency than those of other processes. However, viscosity of $CO_2$ is usually about 1/10 that of oil in the reservoir conditions [8]. As a result, the sweep efficiency of $CO_2$ flooding is lower than efficiency of waterflooding. The water-alternating-gas (WAG) process is suggested by Caudle and Dyes [9] to improve sweep efficiency of $CO_2$ injection. Alternating or coinjection of $CO_2$ and water enhances the recovery of oil. The injected water increases sweep efficiency and stabilizes the gas front. When slugs of $CO_2$ and water are injected into reservoir consecutively, some part of $CO_2$ is dissolved in the oil and reduces the oil viscosity. Thus, the mobility ratio between displacing and displaced fluid is decreased. It becomes favorable condition to control the $CO_2$ breakthrough and improve recovery efficiency.

Another suggested technique which advances sweep efficiency for the heterogeneous reservoir including high permeable thief zones is integrated polymer and $CO_2$ flooding. Generally, polymer flooding is known as effective process when mobility ratio of waterflooding is high, the heterogeneity of reservoir is high, or both of them exist [10]. Polymer flooding is processed by adding polymer into the water to decrease mobility of displacing fluid. Dissolved polymer increases the viscosity of displacing fluid and decreases the effective permeability of aqueous phase through adsorption. High adsorption of polymer through mainly high permeable streaks reduces permeability so that it induces diverting displacing fluid into low permeable zones and increases the oil recovery. However, polymer flooding is not a great way to decrease residual oil saturation. The polymer degradation and shear effect have been problems in application of polymer flood. A substantial amount of polymer is required to reduce the unsuitably high viscosity ratio to a value of approximately one in the heavy oil reservoirs. The significant required number of polymers in such reservoirs leads to high cost [11].

To overcome these problems, such as viscous fingering, poor sweep efficiency, and polymer concentration, integrated EOR method as coupling polymer flooding and $CO_2$ flooding is of importance. It has both advantages of $CO_2$ flooding and polymer flooding, solubility of $CO_2$ injection and mobility control of polymer injection. According to Zhang et al. [11], polymer/gas-alternating-water (PGAW) is combination of these two methods. Majidaie et al. [12] simulated chemically enhanced water-alternating-gas (CWAG) injection in homogeneous reservoir. Li et al. [13] carried out a case study of polymer-alternating-gas (PAG) simulation. However, more research for coupling $CO_2$ flooding and polymer injection is still needed. The previous simulation studies [12, 13] have been carried out in light oil reservoir. Although Zhang et al. [11] assessed its performance considering heavy oil, it is limited with experimental scale. Applications of PAG process in heavy and heterogeneous reservoirs have not been conducted sufficiently. For this reason, specific purpose of this study focused on the simulation of PAG process in field scale heterogeneous reservoir containing heavy oil. To evaluate the effectiveness of PAG process in the model, four processes (waterflooding, continuous $CO_2$ injection, WAG process, and PAG process) are implemented and analyzed with oil recovery factor, WOR, and GOR. In addition, PAG ratio and PAG cycle have been parameterized to maximize the performance of PAG.

# NUMERICAL SIMULATION

## Fluid Modeling

The oil properties of Schrader Bluff and West Sak are referenced for viscous oil modeling. Composition of the oil is reported in Table 1. The portions of intermediate components are small and heavy components are main part. Properties and viscosity data which are used for regression analysis are based on the literature study of Ning et al. (Tables 2 and 3) [14]. Peng and Robinson [15] method is applied to generate PVT data of referenced components.

**Table 1:** Composition of viscous oil

| Component | Mole fraction |
|---|---|
| $CO_2$ | 0.00027 |
| $N_2$ to $C_1$ | 0.30446 |
| $C_2$ to $C_4$ | 0.01018 |
| $C_5$ to $C_7$ | 0.02464 |
| $C_8$ to $C_{12}$ | 0.09672 |
| $C_{13}$ to $C_{19}$ | 0.21201 |
| $C_{20}$ to $C_{30}$ | 0.35172 |
| Total | 1 |

**Table 2:** Properties of reservoir fluid

| Stock tank oil density | 0.953 kg/m³ |
|---|---|
| STO API gravity | 16.9 |
| Gas oil ratio | 32.2 m³/m³ |
| Saturation pressure | 101 atm |

**Table 3:** Viscosity of the reservoir fluid at 24°C

| Pressure (atm) | Viscosity (kg/m·sec) |
|---|---|
| 170 | 0.1411 |
| 136 | 0.1300 |
| 116 | 0.1250 |
| 109 | 0.1225 |

Due to solubility of $CO_2$ into heavy oil, $K$-values are calculated to represent an equilibrium state between components. The definition of $K$-value is the ratio of equilibrium gas component $y_i$ to the equilibrium liquid composition $x_i$ as follows:

$$K_i \equiv \frac{y_i}{x_i}.$$

(1)

$K_i$ is a function of pressure, temperature, and oil composition. $K$-values are calculated by satisfying the fugacity of equilibrium state based on EOS model. For the oil components given in Table 1, $K$-values are estimated on various pressures as depicted in Figure 1.

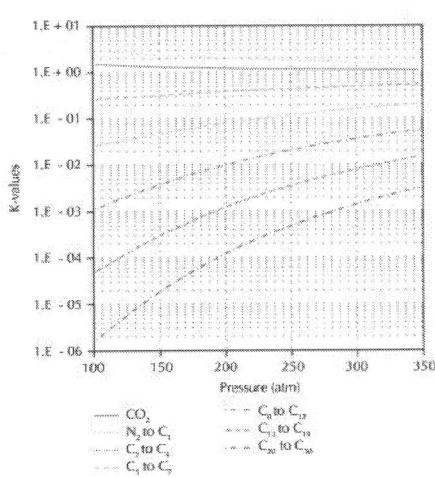

**Figure 1:** Estimations of $K$ values for reservoir oil components at 24°C.

# Hypothetical Reservoir Modeling

The hypothetical reservoir model is assumed as layered model which is discretized into $50 \times 1 \times 10$ grid blocks. Each grid block has dimension of 1.2 m$\times$ 3 m $\times$ 1.5 m (Figure 2). The depth of reservoir is 244 m and reference pressure and temperature at this point are 121 atm and 24° C. The porosity is 30%. Average permeability is $5.3 \times 10^{-8}$ m2 which has Dykstra-Parsons coefficient ($VDP$) representing variation of permeability as 0.75 which is determined by permeability variation [16] as follows:

$$V_{DP} = \frac{k_{50} - k_{84.1}}{k_{50}},$$

(2)

Where $k_{50}$ is permeability value at 50% probability and $k_{84.1}$ is permeability value at 84.1% of the cumulative sample. The range of coefficient varies from 0 to 1. If the heterogeneity of reservoir increases, the value of coefficient approaches to 1. Vertical/horizontal permeability ratio is assumed as 0.1. The initial water saturation is 0.2 and oil saturation is 0.8. Viscosity of water is 0.00045 kg/m $\cdot$ sec and oil and $CO_2$ viscosities are estimated to be about 0.094 kg/m $\cdot$ sec and 0.0001 kg/m $\cdot$ sec, respectively. The viscosity of polymeric solution is 0.022 kg/m·sec at the concentration of 1,000 ppm. Tables 4 and 5 present the input reservoir properties and permeability data used for this simulation. Water is injected during the first year and other processes (waterflooding, continuous $CO_2$ injection, WAG process, and PAG process) are implemented for next 10 years.

**Table 4:** Input data for reservoir simulation

| Parameters | Values |
|---|---|
| Reservoir size (m³) | $60 \times 3 \times 15$ |
| Number of grids | $50 \times 1 \times 10$ |
| Permeability | |
| Average (m²) | $5.3 \times 10^{-8}$ |
| $k_v/k_h$ | 0.1 |

| $V_{DP}$ | 0.75 |
|---|---|
| Porosity | 0.3 |
| Pressure (atm) | 121 |
| Temperature (°C) | 24 |
| Initial saturation | |
| Water | 0.2 |
| Oil | 0.8 |
| Viscosity (kg/m·sec) | |
| Water | 0.00045 |
| Oil | 0.094 |
| $CO_2$ | 0.0001 |
| Polymer 1,000 ppm | 0.022 |

**Table 5:** Permeability data for layered reservoir

| Layer number | Permeability ($10^{-7}$ m²) |
|---|---|
| 1 | 2.4 |
| 2 | 1.6 |
| 3 | 2.6 |
| 4 | 1.3 |
| 5 | 0.89 |
| 6 | 0.69 |
| 7 | 0.44 |
| 8 | 0.20 |
| 9 | 0.15 |
| 10 | 0.08 |

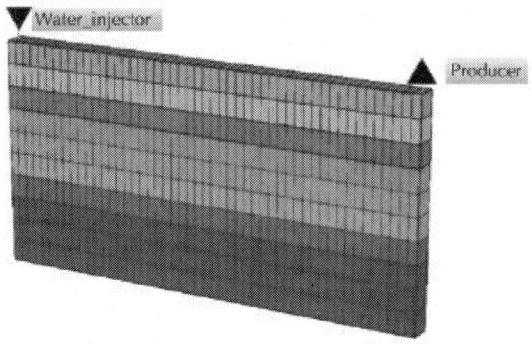

**Figure 2:** Hypothetical model consisting of different permeability layers.

# Mobility Control

The objective of WAG process is originally to aim for the ideal oil recovery system: improvements of macroscopic and microscopic sweep efficiency at once. The injected water (or polymeric solution) is able to control the injected gas mobility as follows:

$$f_w = \frac{k_w/\mu_w}{k_w/\mu_w + k_o/\mu_o + k_g/\mu_g},$$

$$f_g = \frac{k_g/\mu_g}{k_w/\mu_w + k_o/\mu_o + k_g/\mu_g},$$

(3)

Where $f$ is the fractional flow, $k$ is the permeability, and $\mu$ is the viscosity [17].

The oil recovery factor ($R_f$) is determined by microscopic sweep efficiency and the macroscopic sweep efficiency. The macroscopic sweep efficiency can be described by the horizontal and vertical sweep efficiencies. The recovery factor is formulated by

$$R_f = E_v E_h,$$

(4)

Where $E_v$ is the vertical sweep efficiency and $E_h$ is the horizontal sweep efficiency [10].

The mobility ratio (5) [18] affects horizontal sweep efficiency and the vertical sweep efficiency is related to the ratio of viscous to gravity forces (6) [19]. Consider

$$M = \frac{k_{r,\text{displacing fluid}}/\mu_{\text{displacing fluid}}}{k_{r,\text{displaced fluid}}/\mu_{\text{displaced fluid}}},$$

$$\tag{5}$$

$$R_{v/g} = \left(\frac{v\mu_o}{kg\Delta\rho}\right)\left(\frac{L}{h}\right),$$

$$\tag{6}$$

where $M$ is the mobility ratio, $R_{v/g}$ is the viscous/gravity forces ratio, V is Darcy velocity, $\mu_o$ is the oil viscosity, $k$ is the permeability, $g$ is the gravitational acceleration, $\Delta\rho$ is difference in oil and solvent densities, $L$ is distance between wells, and $h$ is height of reservoir.

## Polymer Behavior

The polymer adsorption at reservoir rock could be described by Langmuir-type isotherm [20] such as

$$ad = \frac{(a_1 + a_2 S_b)C_p}{1 + a_3 C_p},$$

$$\tag{7}$$

Where $a_1$, $a_2$, and $a_3$ are coefficients of isothermal Langmuir equation, $S_b$ is the salinity of the brine, and $C_p$ is the mole fraction of polymer. Adsorption is assumed as irreversible process. By means of adsorption, not only more polymer concentration is required to reach target polymer concentration, but also induced reduction of permeability decreases flow capacity [21].

# RESULTS AND DISCUSSION

## Comparison of Processes

This study aims to confirm the effectiveness of PAG process in the heavy oil reservoirs. To examine the performance of various injection processes such as waterflooding, continuous $CO_2$ injection, WAG process, and PAG process, oil recovery factors are compared as depicted in Figure 3. Oil recovery from waterflooding is slightly higher than recovery of $CO_2$ flooding. $CO_2$ flooding has better recovery efficiency than that of waterflooding until the recovery factor reaches 19%. The efficiencies of $CO_2$ flooding and waterflooding are reversed after that point. The reservoir considered in this simulation includes high permeable layer at the top. Gravity overriding effect and early breakthrough mainly occur through the high permeable streak. Figure 4 indicates the gravity overriding effect and $CO_2$ breakthrough after one year of $CO_2$ injection. The breakthrough can develop main $CO_2$ flow path and most of injected $CO_2$ passes through the path. Despite high potential for displacement efficiency, this effect reduces the sweep efficiency in application of $CO_2$ flooding. As this phenomenon makes no more increases in oil recovery after five years of $CO_2$ injection, oil recoveries between waterflooding and $CO_2$ flooding are reversed.

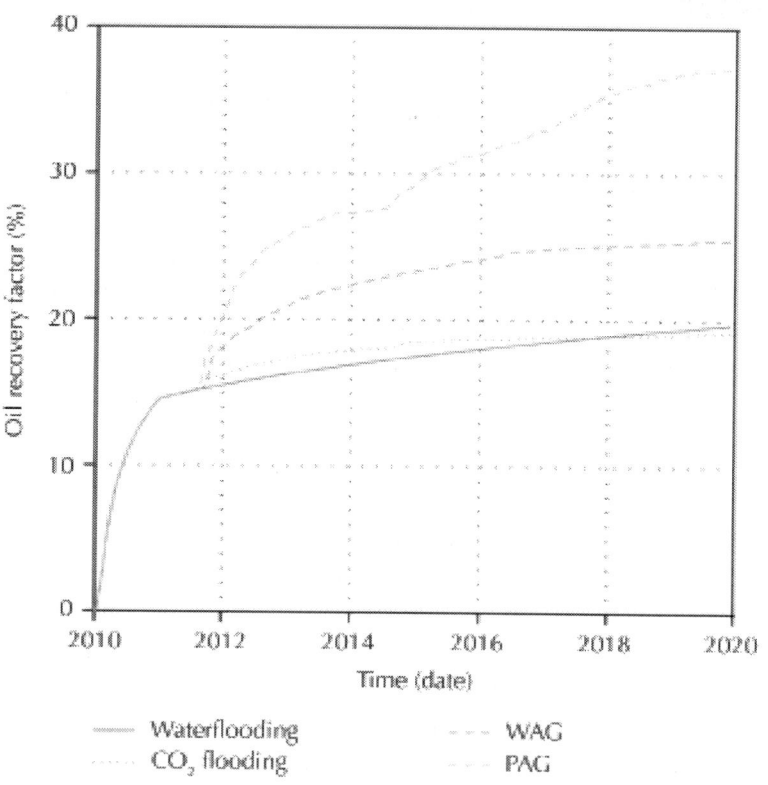

**Figure 3:** Oil recovery factors for different processes.

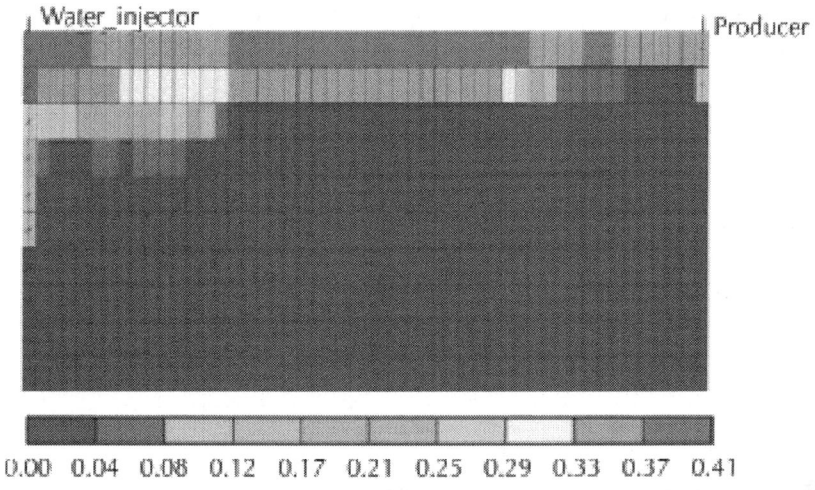

0.00  0.04  0.08  0.12  0.17  0.21  0.25  0.29  0.33  0.37  0.41

**Figure 4:** $CO_2$ mole fraction after one year of $CO_2$ injection in continuous $CO_2$ flooding.

WAG process is implemented and investigated. WAG ratio is set as 1:1 and one cycle period is 180 days, respectively. According to Figure 3, WAG process obtains 26% oil recovery while recovery factors of water flooding and $CO_2$ flooding are less than 20%. This improved oil recovery as much as 6% by application of WAG is reasoned from the increased sweep efficiency and displacement efficiency by applying water and $CO_2$ flooding. A great amount of oil is easily extracted to the producer.

In PAG process, polymeric solution is injected into reservoir instead of water in WAG process. The solution contains 1,000 ppm polymer and it could prove the effect of polymeric solution during PAG process. PAG process achieves the highest oil recovery in Figure 3. PAG process takes 37% recovery factor. The enhancement of oil recovery by PAG method is 89% over water or $CO_2$ flooding and 45% over WAG process. The additional recovery resulted from advance of mobility ratio. The comparison of viscosity between WAG and PAG is reported in Figure 5 which describes viscosity of aqueous phase near the injection well. The viscosity obtained from PAG process continuously increases during the injection period of polymeric solution. The betterment of mobility ratio is due to high viscosity of injected polymeric solution and permeability reduction by adsorption. This improvement is indicated by resistance factor in Figure 6. The resistance factor is a ratio of water mobility to polymeric solution mobility. If the viscosity is increased by polymer injection, resistance factor is increased by reduction of polymer mobility [22]. These processes can alleviate viscosity fingering effect in heterogeneous reservoirs. A channeling due to the permeability heterogeneity of this layered system is a dominant factor to reduce sweep efficiency. Figures 7(a) and 7(b) show that improved mobility ratio in PAG process can mitigate viscosity fingering problem and form a stable front.

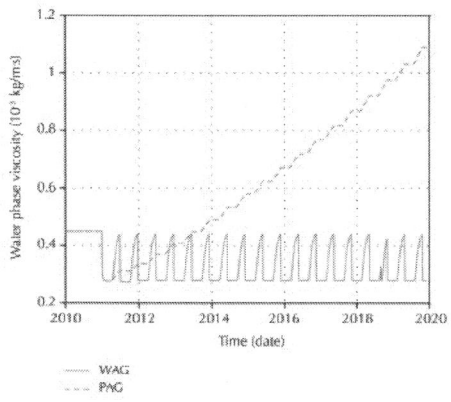

**Figure 5:** Viscosity of aqueous phase in high permeability zone.

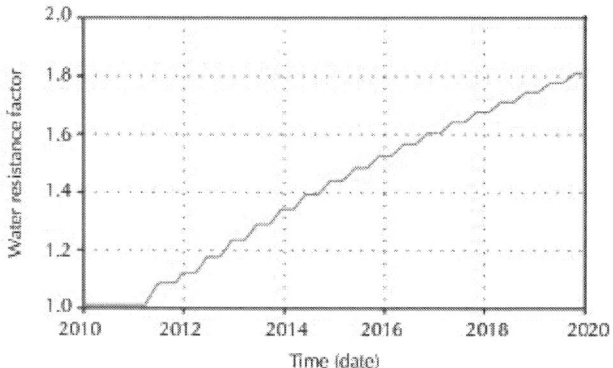

**Figure 6:** Resistance factor for aqueous phase in high permeability zone.

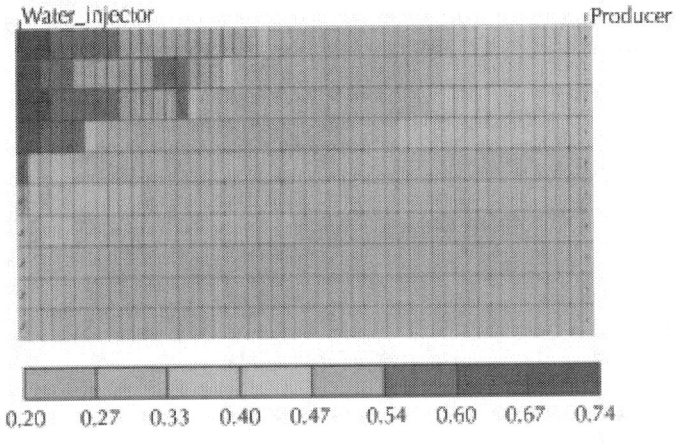

(a)

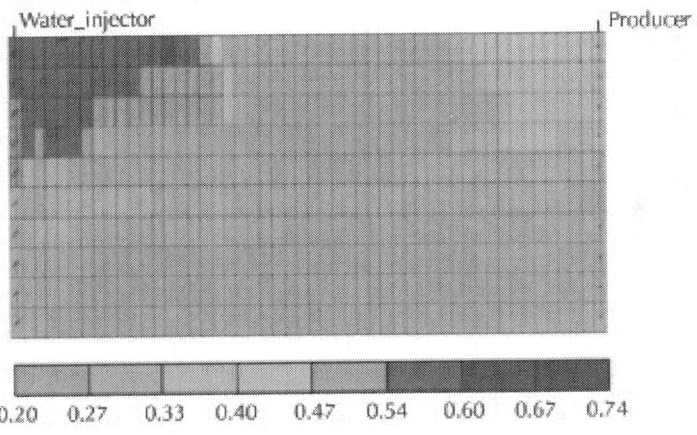

(b)

**Figure 7:** Water saturation after one cycle of WAG and PAG: (a) WAG process and (b) PAG process.

Figures 8 and 9 represent the water-oil ratio (WOR) and cumulative water production from different processes. According to Figure 8, the results signify that the WOR in the case of PAG process is much lower than those of water flooding and WAG process during production period, excepting 2014. At this time, WOR of PAG is sharply increased because produced oil rate reaches almost zero due to the temporary blockage with injected $CO_2$ and polymeric solution (Figure 10). In Figure 9, the PAG process indicates 42% and 12% reduction in cumulative water production compared to the water flooding and WAG process, respectively. These improvements prove effectiveness for polymer injection which has great potential to reduce aqueous phase mobility.

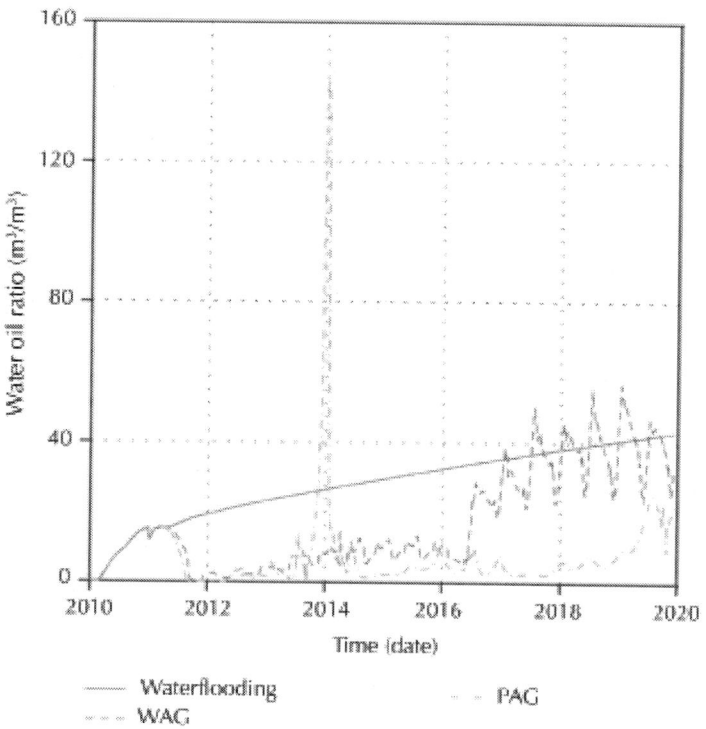

**Figure 8:** Water-oil ratios for different processes.

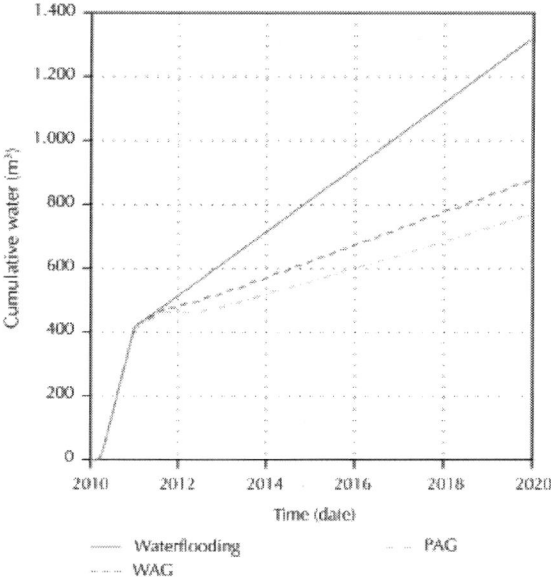

**Figure 9:** Cumulative water productions for different processes.

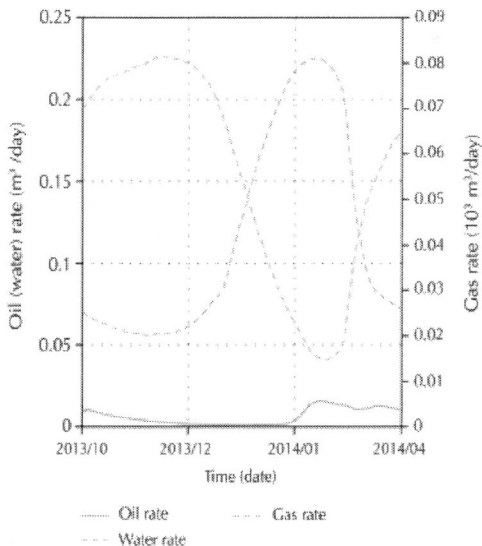

**Figure 10:** Produced oil, water, and $CO_2$ rate of PAG process in 2014.

Figures 11 and 12 describe the gas-oil ratio (GOR) and cumulative gas production. In Figure11, similar problem with WOR existed at 2014. As aforementioned, the same problem at this point results from low oil rate. The amounts of gas productions are $3.4 \times 10^5$ $m^3$ in $CO_2$ flooding and $1.6 \times 10^5$ $m^3$ in WAG and PAG process. PAG process obtains 53% reduction in cumulative gas production compared to $CO_2$ flooding. GOR of PAG process is significantly lower than that of WAG process. PAG process in heterogeneous heavy oil reservoir attains better performance than WAG does.

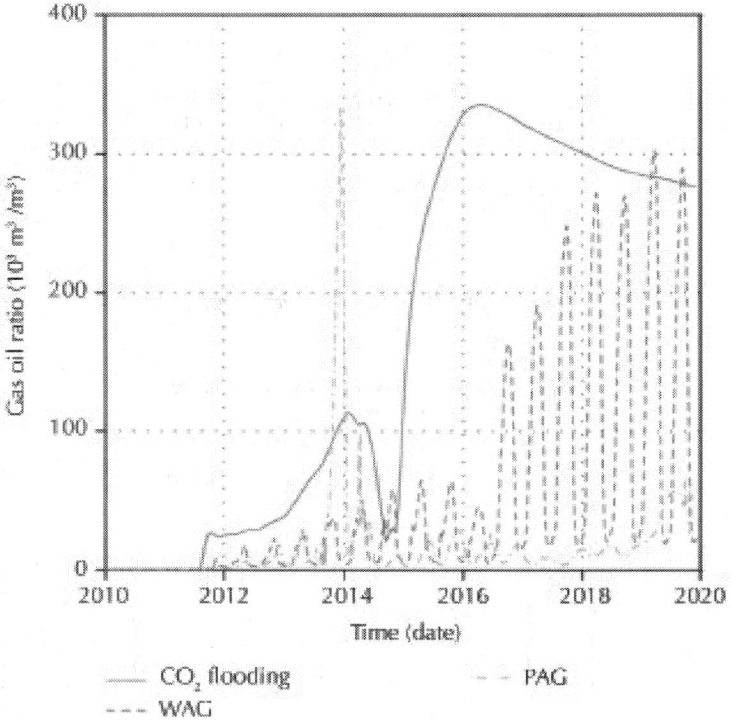

**Figure 11:** Gas-oil ratios for different processes.

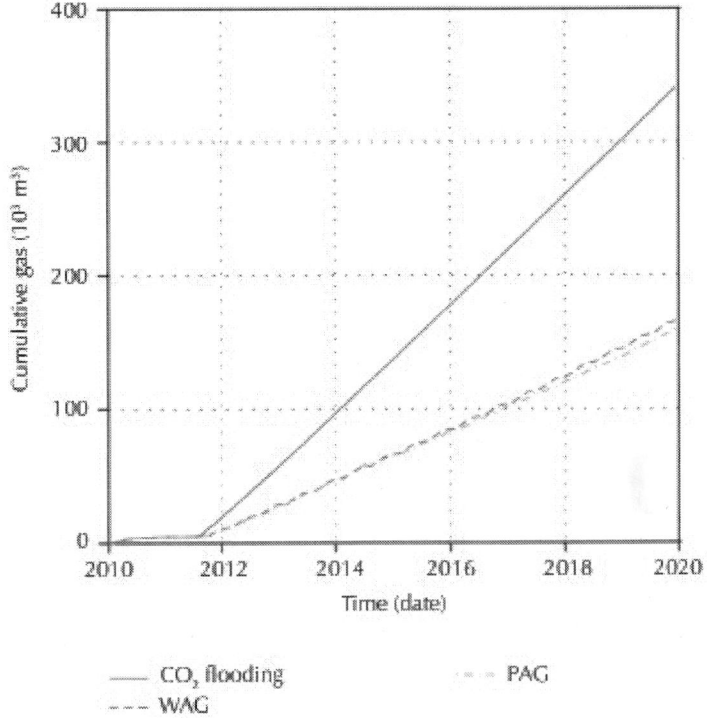

**Figure 12:** Cumulative gas productions for different processes.

Figure 13 depicts the oil saturation distribution for four processes at the end of production. Average oil saturation is 0.64 in water flooding, 0.67 in $CO_2$ flooding, 0.63 in WAG process, and 0.56 in PAG process. In comparison with Figures 13(a), 13(b), and 13(c), Figure 13(d) shows that better recovery efficiency resulted from high sweep efficiency and high displacement efficiency. In PAG process, the reduced permeability contrast due to the preferential adsorption of polymer in relatively high-permeability layers enables water and $CO_2$ to penetrate into low-permeability layers and the recovery efficiency to be increased.

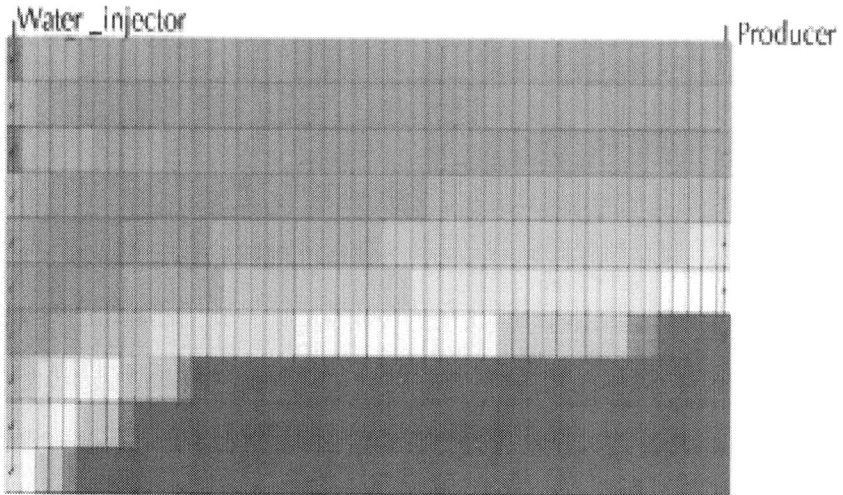

(a)

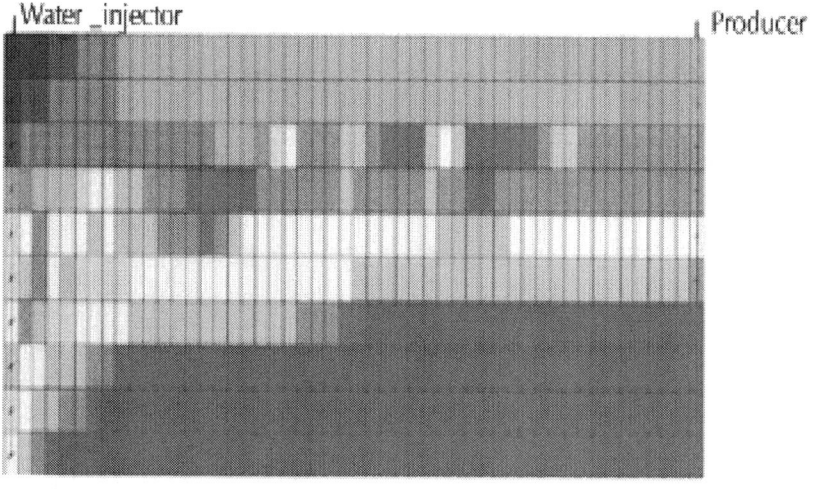

(b)

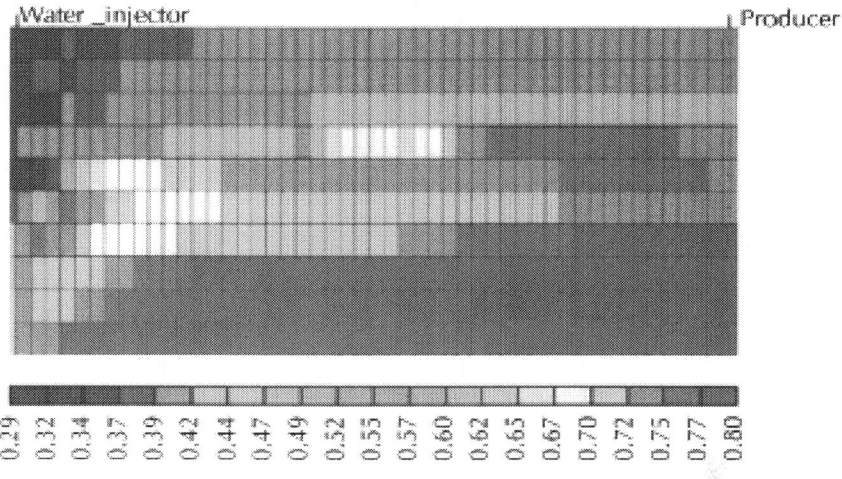

(c)

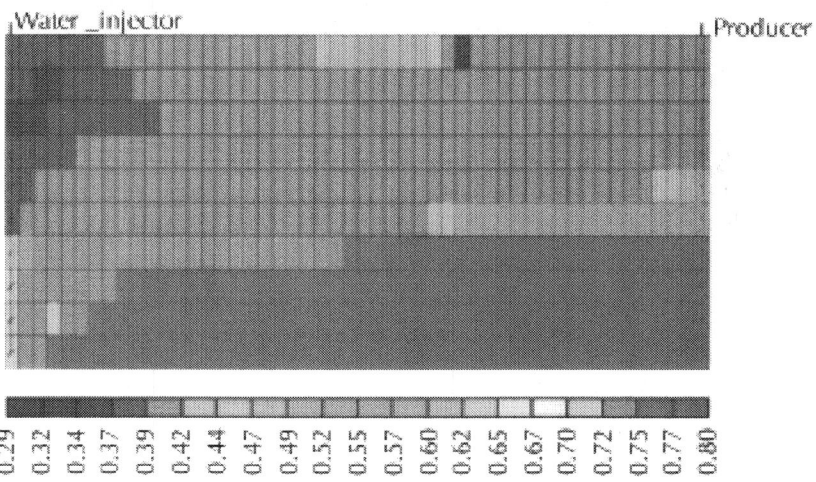

(d)

**Figure 13:** Oil saturations at the end of simulation: (a) waterflooding, (b) continuous $CO_2$ flooding, (c) WAG process, and (d) PAG process.

## PAG Cycle and Ratio

PAG cycle and ratio are general parameters which determine the characteristics of PAG process. The base case is 1:1 PAG ratio and 180 days cycle period. Various PAG cycle periods are applied to compare the oil recovery in the same PAG ratio (1:1). In this PAG process, $CO_2$ is injected first and polymeric solution follows. The results of these processes are shown in Figure 14. Ultimate recoveries are similar for all cases although increasing points of recovery factors are different. If the respective total amounts of injected $CO_2$ and polymeric solution are the same in five cases, they have similar efficiencies of sweep and displacement. These results are well matched with those from previous WAG simulation study [23].

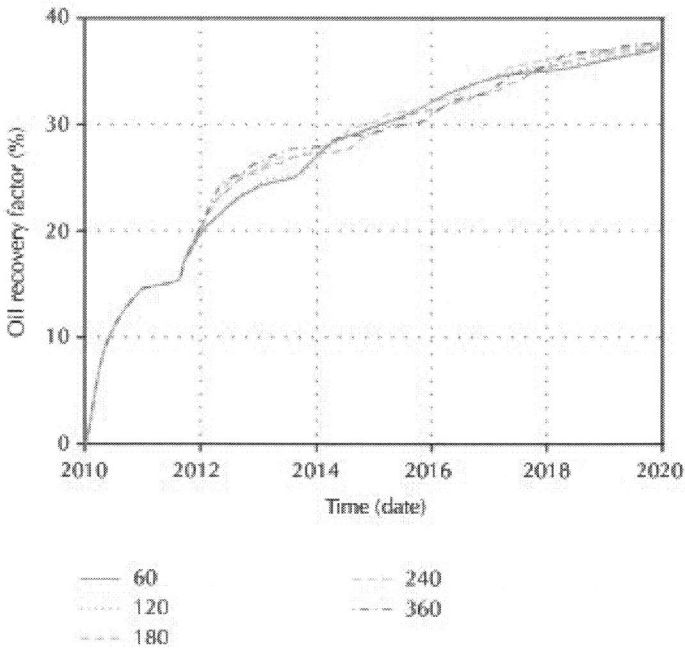

Figure 14: Oil recovery factors at different WAG cycle periods.

Figure 15 is the result of oil recoveries for different PAG ratio processes. Oil recovery factor of 2:1 PAG ratio is 3% larger than that of 1:1 PAG ratio. As a result, injection of more polymeric solution has advantage for oil recovery by increased sweep efficiency. However, too much polymer injection could reduce the oil recovery because mobility of polymer is low and polymer does not reduce residual oil saturation (1:5 PAG ratio case).

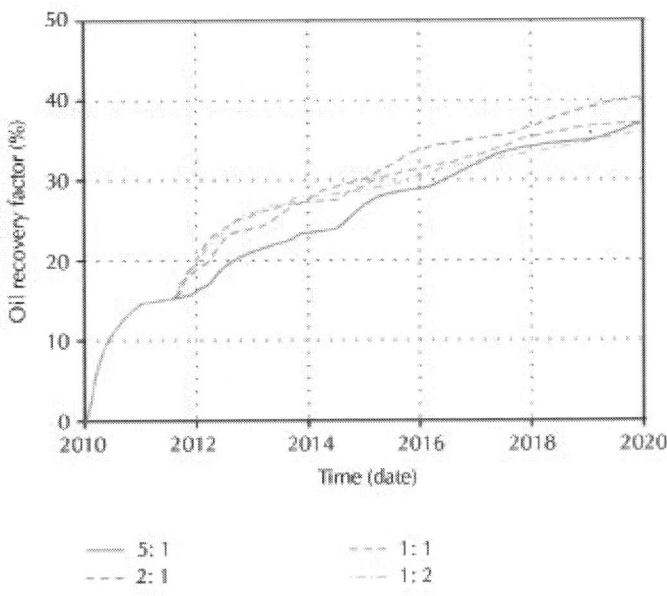

**Figure 15:** Oil recovery factors at different WAG ratio cases.

# CONCLUSIONS

The results of numerical simulation on the flow of polymeric solution with $CO_2$ in heavy oil reservoir were analyzed. The main challenges to reduce oil recovery are high viscosity of heavy oil and high heterogeneity of reservoir. The polymer-alternating-gas process showed significant advance of recovery efficiency compared with other processes.

- By the control of mobility ratio, the PAG process has better sweep efficiency than those of other processes. The PAG process represented the highest oil recovery factor by 37%. It was 89% higher than results of waterflooding or $CO_2$ flooding and 45% higher than consequence of WAG process.

- In heterogeneous heavy oil reservoir, water and $CO_2$ breakthroughs are key factors to reduce the oil recovery in WAG process. In comparison to WAG process, PAG process would decrease the WOR by 12%. Moreover, GOR in PAG process was maintained below GOR of WAG method.

- The cycle time of PAG process did not affect the recovery performance. However, 2:1 PAG ratio could improve the oil recovery factor by about 3% over the base case.

# ACKNOWLEDGMENTS

This work was supported by the Energy Efficiency & Resources Core Technology Program of the Korea Institute of Energy Technology Evaluation and Planning (KETEP) granted financial resource from the Ministry of Trade, Industry & Energy, Republic of Korea (no. 20122010200060).

# REFERENCES

1.    T. B. Reid and H. J. Robinson, "Lick creek meakin sand unit immiscible $CO_2$ waterflood project," Journal of Petroleum Technology, vol. 33, no. 9, pp. 1723–1729, 1981.

2.    K. Kantar, D. Karaoguz, K. Issever, and L. Varana, "Design concepts of a heavy-oil recovery process by an immiscible $CO_2$ application," Journal of Petroleum Technology, vol. 37, no. 2, pp. 275–283, 1985.

3.    W. E. Kerr, "Retlaw upper mannvile "V" pool unit experimental carbon dioxide flood," Journal of Canadian Petroleum Technology, vol. 24, no. 1, pp. 275–283, 1985.

4.    A. Spivak, W. H. Garrison, and J. P. Ngugen, "Review of an immiscible $CO_2$ project, tar zone, fault block V, Wilmington field, California," SPE Reservoir Engineering, vol. 5, no. 2, pp. 155–162, 1990.

5.    W. S. Fong, R. W. Tang, A. S. Emanuel, P. J. Sabat, and D. A. Lambertz, "EOR for California diatomites. $CO_2$, flue gas and water corefloods, and computer simulations," in Proceedings of

the Western Regional Meeting, pp. 159–170, Bakersfield, Calif, USA, April 1992

6.   K. Issever, N. A. Pamir, and A. Tirek, "Performance of a heavy-oil field under $CO_2$ injection, Bati Raman, Turkey," SPE Reservoir Engineering, vol. 8, no. 4, pp. 256–260, 1993.

7.   T. D. Ma and G. K. Youngren, "Performance of Immiscible Water-Alternating-Gas (IWAG) Injection at Kuparuk River Unit, North Slope, Alaska," in Proceedings of the SPE Annual Technical Conference and Exhibition, pp. 415–420, New Orleans, La, USA, September 1994

8.   D. N. Rao, S. C. Ayirala, M. M. Kulkarni, and A. P. Sharma, "Development of gas assisted gravity drainage (GAGD) process for improved light oil recovery," inProceedings of the SPE/DOE Symposium on Improved Oil Recovery, Tulsa, Okla, USA, April 2004.

9.   B. H. Caudle and A. B. Dyes, "Improving Miscible displacement by gas-water injection,"Petroleum Transactions of the AIME, vol. 213, pp. 281–284, 1958.

10.   L. W. Lake, Enhanced Oil Recovery, Society of Petroleum Engineers, Richardson, Tex, USA, 2010.

11.   Y. Zhang, S. Huang, and P. Luo, "Coupling immiscible $CO_2$ technology and polymer injection to maximize EOR performance for heavy oils," Journal of Canadian Petroleum Technology, vol. 49, no. 5, pp. 27–33, 2010

12.   S. Majidaie, A. Khanifar, M. Onur, and I. M. Tan, "A simulation study of chemically enhanced water alternating gas (CWAG) injection," in Proceedings of the SPE EOR Conference at Oil and Gas West Asia, pp. 242–250, Muscat, Oman, April 2012.

13.   W. Li, Z. Dong, J. Sun, and D. S. Schechter, "Polymer-alternating-gas simulation: A Case Study," in Proceedings of the SPE EOR Conference at Oil and Gas West Asia, Muscat, Oman, March 2014.

14.   S. X. Ning, B. S. Jhaveri, N. Jia, B. Chambers, and J. Gao, "Viscosity reduction EOR with $CO_2$ & enriched $CO_2$ to improve recovery of Alaska North Slope viscous oils," inProceedings of the SPE Western North American Region Meeting, Anchorage, Alaska, USA, 11.

15. D. Y. Peng and D. B. Robinson, "A new two-constant equation of state," Industrial and Engineering Chemistry Fundamentals, vol. 15, no. 1, pp. 59–64, 1976.

16. H. Dykstra and R. L. Rarsons, "The prediction of oil recovery by waterflood," inSecondary Recovery of Oil in the United States, pp. 160–175, 1950.

17. S. E. Buckley and M. C. Leverett, "Mechanism of fluid displacement in sands,"Transactions of AIME, vol. 146, no. 1, pp. 107–116, 1942.

18. F. F. Craig Jr., T. M. Geffen, and R. A. Morse, "Oil recovery performance of pattern gas or water injection operations from model tests," Transactions of AIME, vol. 204, pp. 7–15, 1955.

19. F. I. Stalkup Jr., Miscible Displacement, Society of Petroleum, Richardson, Tex, USA, 1983.

20. Computer Modelling Group Ltd, "STARS: Advanced Process and Thermal Reservoir Simulator," Calgary, Canada, 2012.

21. J. J. Sheng, Modern Chemical Enhanced Oil Recovery: Theory and Practice, Gulf Professional Publishing, Burlington, Mass, USA, 2010.

22. R. R. Jennings, J. H. Rogers, and T. J. West, "Factors influencing mobility control by polymeric solutions," Journal of Petroleum Technology, vol. 23, no. 3, pp. 391–401, 1971.

23. E. L. Ligero, S. F. Mello, E. O. Muñoz Mazo, and D. J. Schiozer, "An approach to oil production forecasting in a WAG process using natural $CO_2$," in Proceedings of the SPETT Energy Conference and Exhibition, pp. 109–118, Port of Spain, Trinidad and Tobago, June 2012.

# Chapter 5

# A Novel Energy-efficient Pyrolysis Process: Self-pyrolysis of Oil Shale Triggered by Topochemical Heat in a Horizontal Fixed Bed

You-Hong Sun[1], Feng-Tian Bai[1], Xiao-Shu Lü, Qiang Li[1, 2], Yu-Min Liu[1], Ming-Yi Guo[1], Wei Guo[1], and Bao-Chang Liu[1]

[1]College of Construction Engineering, Jilin University, Changchun 130021, PR China
[2]Department of Civil and Structural Engineering, School of Engineering, Aalto University, Espoo, Finland

## ABSTRACT

This paper proposes a novel energy-efficient oil shale pyrolysis process triggered by a topochemical reaction that can be applied in horizontal oil shale formations. The process starts by feeding preheated air to oil shale to initiate a topochemical reaction and the onset of self-pyrolysis. As the temperature in the virgin oil shale increases (to 250–300°C), the hot air can be replaced by ambient-temperature air, allowing heat to be released by internal topochemical reactions to complete the pyrolysis. The propagation of fronts formed in this process, the temperature evolution, and the reaction mechanism of oil shale pyrolysis in porous media are discussed and compared with those in a traditional oxygen-free process. The results show that the self-pyrolysis of oil shale can be achieved with the proposed method without any need for external heat. The results also verify that fractured oil shale may be more suitable for underground retorting. Moreover, the gas and liquid products from this method were characterised, and a highly instrumented experimental device designed specifically for this process is described. This study can serve as a reference for new ideas on oil shale in situ pyrolysis processes.

# INTRODUCTION

The tremendous growth in global energy demand and the rapid depletion of conventional oil resources have created a demand for alternative energy sources. Oil shale[1, 2, 3, 4], an organic-rich petroleum source rock, has a high content of solid insoluble kerogen that can release a petroleum-like liquid, that is, shale oil, and provide secure access to transportation fuels[5, 6]. The huge reserves of oil shale, approximately 35% of the world's total energy reserves[7], make it commercially viable. Accordingly, scientists have been attempting for decades to develop environmentally responsible methods of exploiting oil shale[8, 9, 10, 11, 12, 13, 14].

Oil shale can be used in various ways[15, 16, 17, 18, 19, 20], from electrical energy via direct combustion to a wide range of petrochemical products (including shale oil and other liquid fuels) via the pyrolysis of kerogen. Two primary types of processes for oil shale retorting have been developed: ex situ and in situ processes. Ex situ (aboveground) processes[16, 20], such as retorts of the Kiviter, Petrosix, ATP, Tosco-II and Fushun-type, are the primary extraction methods. However, such methods suffer critical problems[8, 16, 17, 21, 22], including potentially harmful semicoke waste, large land use, and gas pollution. Additionally, ex situ processes are unsuitable for low-grade oil shale buried in deep formations because of mining difficulties and negative economic impacts. In situ processes[10, 11], such as ICP, Electrofrac™, Chevron's technology, and the Occidental Modified in-situ (MIS) process, have recently attracted great attention, without regard to the heating modes and environmental issues related to these methods. The ICP and Electrofrac™ processes use electricity to heat the surrounding shale deposit as a physical heat treatment, consuming significant energy[10]. Chevron's technology uses heated and pressurised carbon dioxide to heat the oil shale, which also requires large quantities of water and damages the environment. The MIS method induces combustion air to permeate and burn oil shale underground after mining 20% of the oil shale and fracturing the rest to create a void space of 20–25%; this method was conducted in situ in 1972 at Logan Wash, Colorado[10].

We recently introduced the oil shale topochemical reaction[23] as a low-energy shale oil recovery method that is triggered by the topochemical reaction between oil shale and a limited amount of air to drive the spontaneous pyrolysis of oil shale in the absence of an external heat supply (denoted as a self-pyrolysis process). In this method, the pyrolysis represents a chemical-enhanced heating process rather than a physical heating process, like ICP and Electrofrac™, or a complete combustion process. Moreover, the shale oil product still consists primarily of hydrocarbons, like the oils obtained from traditional processes. The proposed technique[23] has the potential to consume much less energy, making it more suitable for underground retorting. However, neither the heat transfer mechanisms of the proposed method nor the importance of process design in realising energy savings has been elucidated. Furthermore, previous experiments have been limited to data obtained from a tube furnace for the analysis of products, with little data on heat transport properties. Results cannot therefore be directly applied to energy-efficient pyrolysis solutions. The purpose of this paper is to address these gaps. We aim to provide new insights into the mechanism of the topochemical-induced thermal process of oil shale conversion to gain a novel understanding of the self-pyrolysis process of oil shale and demonstrate a more feasible energy-efficient approach through a laboratory-scale process.

Oil shale generally exhibits poor thermal conductivity and permeability and is therefore very inefficient at transmitting heat and gas. For ex situ retorting, shale is always crushed into smaller particles (commonly < 100 mm) to increase its specific surface and contact areas. For in situ methods, porosity and permeability are two important factors that enable the transfer of heat to the shale within the geologic formation[14]. Many methods[14, 24, 25], including the use of explosives, hydraulic fracturing, and horizontal drilling, can be used to rubblize the formation to improve the stratum permeability and increase the reaction area. In these experiments, various sizes (2–100 mm) of oil shale were introduced into a fixed bed to form a heterogeneous system to mimic different permeabilities and porosities of shale in the field.

In the present study, a self-pyrolysis retorting process of oil shale with a low energy input using high- and normal-temperature air in sequence as carrier gases is described. A highly instrumented fixed bed (Figure 1a) simulating the horizontal stratum was developed to investigate this novel pyrolysis in a porous medium under controlled conditions. The propagation of fronts formed in this process, the temperature evolution, and the reaction mechanism of oil shale pyrolysis in porous media are discussed and compared with those in a traditional oxygen-free process. Two different circular gas pipelines (Figures 1b, c) were specifically designed to simulate the in situ hydraulic fracturing and horizontal drilling modes. The products from this method were also characterised. We anticipate that this study can serve as a starting point to initiate new approaches to oil shale in situ pyrolysis processes.

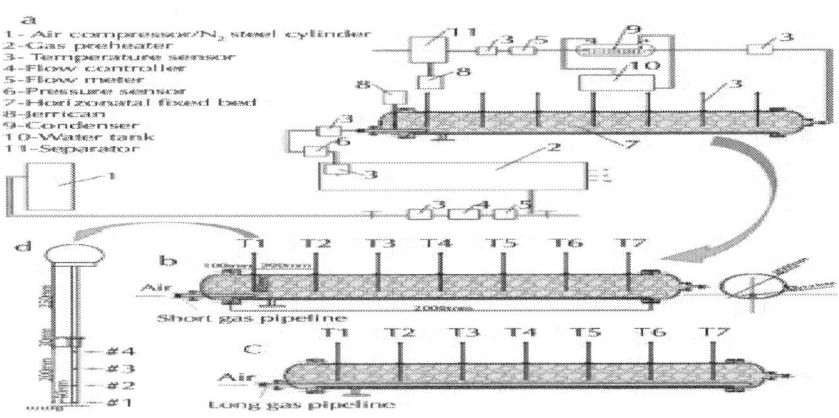

**Figure 1:** Schematic of the experimental device. (a), the overall layout of the device. (b), the fixed bed with short gas pipeline. (c), the fixed bed with long gas pipeline. (d), the temperature sensor.

# RESULTS

## Thermogravimetric (TG) Analysis, Differential Thermogravimetric (DTG) Analysis and Differential Scanning Calorimetry (DSC)

The results of the TG, DTG, and DSC experiments on Huadian (HD) oil shale under both air and inert ($N_2$) atmospheres are shown in Figure 2. The experiments were repeated several times and exhibited good reproducibility. The stages are as follows.

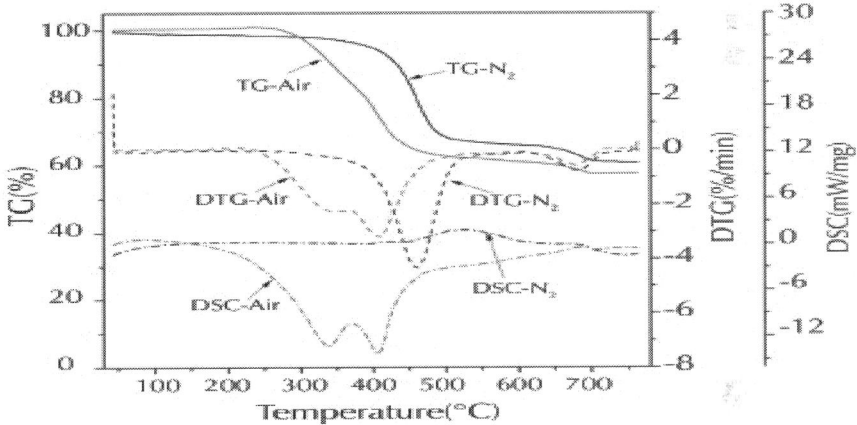

**Figure 2:** TG/DTG/DSC results for virgin HD oil shale under air and under nitrogen.

- At temperatures less than 200°C, a small mass loss is observed, due primarily to the evaporation of water, including adsorbed and interlayer water, from clay minerals.
- Under an air atmosphere, a slight rebound is observed at approximately 300°C because of the absorption of ambient gas[17]; this result differs from the results obtained under an $N_2$ atmosphere.
- In the temperature range from 300°C to 550°C, a major mass loss is observed. This stage is attributed primarily to the decomposition of kerogen into volatiles. Under an air atmosphere, the reaction is

more rapid, and two peaks are observed in both the DTG and DSC curves. The first peak, referred to as the low-temperature oxidation peak, is observed at 336°C and is due to the oxidation reaction of light hydrocarbons formed by cracking. The second peak, referred to as the high-temperature oxidation peak, appears at 405°C and originates from the oxidation of heavy hydrocarbons, fixed carbon and possibly other components. The two processes together gave off 10.693 MJ kg$^{-1}$ of heat, which is consistent with the proximate analysis results. However, these two exothermic peaks did not appear when the experiment was performed under $N_2$, and the total endothermic heat was 1.191 MJ kg$^{-1}$ under an $N_2$ atmosphere. These results reveal that the exothermic reaction heat from oil shale oxidation is approximately nine times the heat required for oil shale pyrolysis. Therefore, the introduction of a reasonable amount of oxygen to partially oxidise the oil shale and release heat for the surrounding oil shale pyrolysis would facilitate the production of petrochemical goods from oil shale.

- At temperatures greater than 600°C, a final mass loss is observed because of the thermal decomposition of carbonates and clay minerals; the endothermic peak appears in two experiments. The endothermic quantity under air was 49.4% of that under $N_2$.

## Evolution of Temperature in Oil Shale Self-Pyrolysis

The gas pipeline shown in Figure 1b was used for the oil shale self-pyrolysis experiments, and the rubblized oil shale was introduced into the bed to study its heat transfer process. An air compressor was used in these experiments to provide air. Air was preheated and maintained at 400–500°C with a flow rate of 16 m$^3$ h$^{-1}$ before being fed into the fixed bed. The temperatures at different locations inside the bed are reported in Figure 3a. For sensor T1, which was positioned close to the gas entrance, the temperature of T1-#1 (sensor location 1; sensor number 1) slowly increased to 220°C in the first 27 min before rapidly increasing to 600°C within 3 min, triggering a topochemical reaction in the oil shale. The temperature successively increased along the vertical direction of the gas pipeline, indicating that the front propagated as a vertical surface. The temperature propagation was consistent with the gas flow direction, i.e., co-current transmit (the reaction front and feeding gas take place on the same side of the bed), spreading from left to right along the horizontal direction of the gas pipeline.

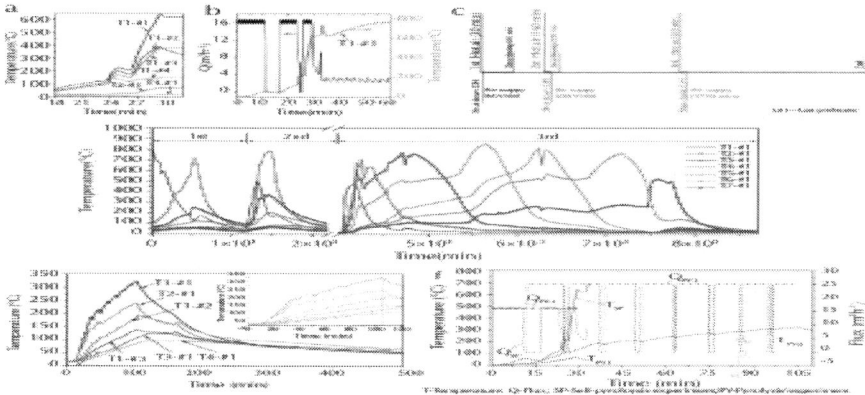

**Figure 3:** The temperature evolution curves and experimental timeline. (a), Temperature evolution of the thermocouples in the self-pyrolysis experiment between 18 and 33 min. (b), Interaction between temperature evolution and gas flow. (c), Timeline of time-interval trials. (d), Temperature evolution of the #1 thermocouples during the time-interval of the self-pyrolysis trials. (e), Temperature evolution of the thermocouples in the pyrolysis process (25 m³ h⁻¹). (f), Temperature evolution of thermocouple T1-#1 in different experimental processes.

Figure 3b clearly shows that the temperature is affected by the fluctuation in the gas flow and that it is characterised by a positive relationship. Temperatures increased as the gas flow increased and rapidly decreased as the flow decreased in all stages of the reaction. After the topochemical reaction is triggered, the temperature can reach 800°C or higher if the flow is excessive. As shown in Figure 3b, the temperature of T1-#1 quickly increased to 800°C in 33 min. This continuous high-temperature condition can consume a large amount of organic matter and is not beneficial to the extraction of shale oil. Conversely, the reactions will stop and the oil shale will cool if the gas source is cut off. A reasonable flow can maintain a steadily propagating front with a slowly increasing temperature, and the oil shale can be pyrolysed smoothly without external heat provision. This lack of need of external heat is due to the topochemical reaction between the oil shale and the oxygen from the air, which spontaneously increases the temperature of the oil shale in the beginning before driving the continuous pyrolysis of kerogen and accomplishing the retorting of the oil shale.

# Time-Interval Trials of Oil Shale Self-Pyrolysis

As shown in Figure 3c, three time-interval trials of self-pyrolysis were run to verify the feasibility and explore the mechanism of the proposed method. Hot air was firstly fed into the fixed bed. When the temperature of T1 increased to 250–300°C, the gas preheating was stopped and the hot gas was replaced by ambient-temperature air. The gas flow was adjusted simultaneously. After a given period of time, the gas source was shut off and the fixed bed was allowed to cool naturally. This trial process was repeated twice. The interval of the first discontinuation was 630 min, and the second (2520 min) was longer than the first to allow the fixed bed to completely cool.

As shown in Figure 3c, the pyrolysis was activated by hot air sweeping for 30 min in the first trials, whereas, in the second and third trials, much longer times were required to induce the reaction. The reason for the longer times was that, during the first trial, the oil shale near sensor T1 was pyrolysed and the front propagated and reached T2 (see Figure 3d), which led to the reduction of the thermal parameters of the residues (Table 1); this reduction resulted in more exchange heat being consumed by the residues during the initial stage of the second and third trials. However, the self-pyrolysis reaction of oil shale is easily triggered, and once started, it can spontaneously and smoothly propagate in the absence of any external heat supply.

**Table 1:** Physical properties of the HD oil shale

|  | Virgin oil shale | Spent shale |
|---|---|---|
| Density (kg m−3) | 1537 | 790 |
| Proximate analysis (wt.%, ad a) |  |  |
| Volatiles | 34.46 | 1.76 |
| Fixed carbon | 3.75 | 0.22 |

| Ash | 61.80 | 98.46 |
|---|---|---|
| Moisture (as received) | 5.53 | 0.59 |
| Calorific value (MJ kg−1) | 11.21 | 0.42 |
| Ultimate analysis (wt.%, ad) | | |
| C | 26.38 | 0.79 |
| H | 4.16 | 0.17 |
| N | 0.31 | 0 |
| S | 1.04 | 2.21 |
| Fischer assay analysis (wt.%, ad) | | |
| Shale oil | 19.69 | 0.87 |
| Gas | 6.77 | 1.15 |
| Water | 6.92 | 1.8 |
| Residue | 66.62 | 96.18 |
| Thermal properties | | |
| Thermal conductivity (W m−1K−1) | 0.68 | 0.17 |
| Thermal diffusivity (mm2 s−1) | 0.50 | 0.68 |
| Specific heat (MJ m−3K−1) | 1.36 | 0.26 |
| Porosity (%,ad) | 1.44 | 58.98 |
| Permeability (mD, ad) | 0.002 | 0.395 |

Ad: air dry basis.

The peak temperature of the #1 thermocouples reached in sequence, as shown in Figure 3d, again reveals that, for the bed equipped with a short pipeline, the pyrolysis fronts caused by the topochemical reaction were co-current with the feeding of air. The temperature fluctuation of the sensors shown in Figure 3d might be explained by the complex flow developed in the bed, which was caused by the conversion of solid kerogen into liquid oil and then into gas[26, 27], generating overpressure in the bed. The four thermocouples of all sensors reached their peak temperatures at approximately the same time, indicating that sharp fronts were generated during propagation. For all sensors except T1, the #3 thermocouple was always the first one to register a temperature increase and the #1 thermocouple always registered the highest temperature. Such a phenomenon could be caused by the flow's upward mobility and the significant heat loss at the wall of the bed.

# Evolution of Temperature in an Oxygen-Free Pyrolysis Process

In a traditional oxygen-free oil shale pyrolysis experiment, the gas was replaced by hot nitrogen at two different speeds of 16 and 25 $m^3$ $h^{-1}$. Figure 3e shows the results for a gas flow of 25 $m^3$ $h^{-1}$. During this experiment, ten cylinders (4 $m^3$ cylinder$^{-1}$) of nitrogen were used; the small zigzags in the curves are associated with the replacement of the nitrogen cylinders. The nitrogen gas had been preheated before being fed into the fixed bed within 105 min; however, the highest temperature in this bed was only 325°C. When the nitrogen gas was no longer preheated, the fixed-bed temperature began to plunge, indicating that nitrogen was just a heat carrier. Moreover, the fixed bed could only reach a temperature of 76°C in 30 min when the gas flow rate was adjusted to 16 $m^3$ $h^{-1}$, as shown in Figure 3f.

# The Structure of the Reaction Fronts for Different Gas Pipelines

Horizontal drilling connecting two boreholes is regarded as a suitable method for oil shale in situ retorting, especially for a thinner shale layer, because it can increase the interaction between the gas and the oil shale layer, which is also beneficial to the transmission of the shale oil. In the present study, the gas channel shown in Figure 1c was designed to mimic the in situ horizontal drilling mode. With a long gas pipeline, the first sensor to register an increase was T7-#1 (located at the outlet of the bed), followed successively by T6-#1, T7-#2, T6-#2, T7-#3, and T6-#3. The highest entrance temperature near T1 was less than 150°C. The front propagation in the long pipeline, from right to left, was completely opposite to that of the short pipeline because it followed a counter-current mode (with the reaction front and feed of gas occurring on opposite sides of the bed). Moreover, the temperature of T4 increased slowly, and that of T3 required a longer time to increase.

# Analysis of the Produced Gas and Liquid Oil and the Solid Residue

In this part, the products of the oil shale self-pyrolysis were analysed. During each experiment, a total of 110–130 kg of oil shale was fed into the bed. The shale oil, water and residue collected from three separate and repeated experiments are reported in Figure 4a. Compared with the Fischer analysis of both virgin and spent oil shale results in Table 1, only a small amount of shale oil (an average of 1.9%) was consumed in the oil shale self-pyrolysis process with air. The amounts of water and gas increased and the amount of residue decreased because of the oxidation of fixed carbon and volatile substances. The main components of the products were analysed as follows.

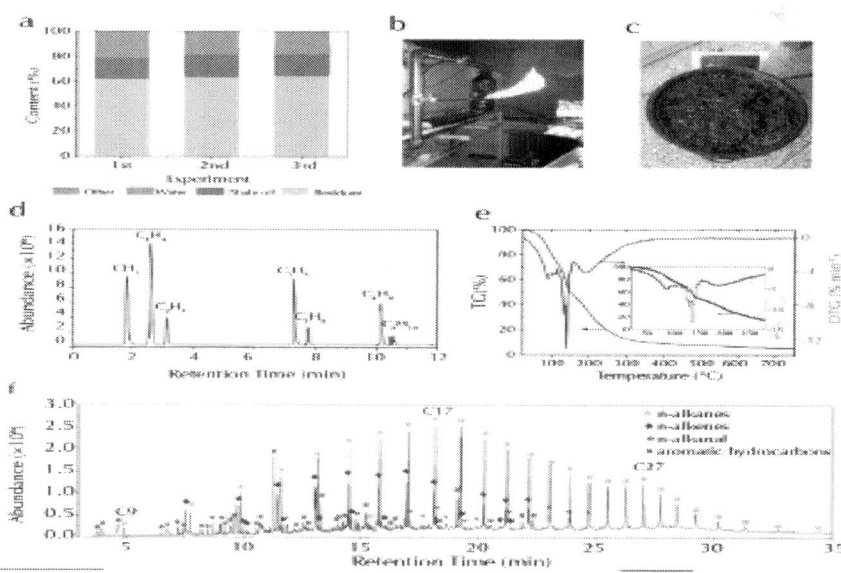

**Figure 4:** The product analysis. (a), the yield of products from the self-pyrolysis experiments. (b), Shale gas ignition tests. (c), Shale oil collected during the experiments. (d), Chromatogram of gas collected from an experiment. (e), TG/DTG analysis of the shale oil. (f), GC-MS spectrum of shale oil collected from the experiment.

A large quantity of pyrolysed gas was generated with the increase in temperature. The colour of the vapour changed to dense-white as the temperature increased, and a large amount of shale gas discharged even when the air was no longer supplied to the fixed bed. The gas exhibited a pungent odour and was flammable (Figure 4b). The shale gases were determined by gas chromatography (GC) (Figure 4d). The chromatograms revealed that the major hydrocarbon components were methane, ethylene, ethane, propylene, propane, butene, and butane.

Oil was gradually expelled from the bed into a jerrican during the experiment. When the experiment was stopped midway through, no liquid oil was observed in the porous medium. Like petroleum, shale oil (Figure 4c) is a black-brown material with an irritating odour. In addition to its use as a transport fuel[5, 6], shale oil can also serve as a virgin material for the production of numerous different value-added products[28, 29], such as rubber softeners, chemical intermediates, and asphalt additives.

TG analysis under $N_2$ (Figure 4e) was carried out primarily to study the characteristics of the boiling point/range of the oil samples. The mass loss between the ambient temperature and 400°C was primarily due to the distillation of the volatile hydrocarbons at 20–280°C and of the low-molecular-weight hydrocarbons at 280–400°C; the mass loss between 400 and 500°C was due to a combination of distillation and thermal cracking of medium-molecular-weight hydrocarbons[30, 31,32]. In this study, the mass loss at 20–280°C was 85%, indicating that the shale oil obtained from our experiments was primarily volatile hydrocarbons. A small amount of residue, approximately 7%, remained in the pan even after the sample was to above 600°C, similar to the residue observed for crude oil[30]. The GC results for the liquids (Figure 4f) show that numerous hydrocarbons and derivatives were present in the collected oils, mainly as aliphatic (63.66%), aromatic (22.80%), and heteroatom (13.54%) compounds. Among these hydrocarbons, aromatic hydrocarbons were primarily produced in the first 16 min and aliphatic hydrocarbons were dominated by a bimodal distribution of n-alkane and n-alkene doublets that extended up to $C_{34}$.

# DISCUSSION

A low-energy-cost strategy for oil shale pyrolysis has been developed to optimise oil shale retorting technology. Different from the typical methods that use hot nitrogen as the carrier gas, the new approach is based on chemical-enhanced heating derived from the topochemical reaction inside the oil shale under preheated air. Thermal analysis shows that, after the topochemical reaction has been triggered by preheated air, it can provide sufficient heat for the subsequent pyrolysis without external heat consumption. Thus, it is a highly energy-efficient self-pyrolysis process compared to the conventional methods. Moreover, an experimental device was designed and constructed to perform this low-energy-cost oil shale pyrolysis in the lab.

The aforementioned results provide new insights into the mechanisms of topochemically induced heating for oil shale pyrolysis, which are critically important for determining the energy required for completing the pyrolysis without external heat. As shown in Figure 5, multiple zones were identified in the bed: the spent shale zone, the retorting zone, and the preheating zone. In the spent shale zone, the heat generated from topochemical reactions between oxygen, water and fixed carbon or that generated from produced gas[23] is applied to the subsequent oil shale pyrolysis in the retorting zone. In the retorting zone, the organics or kerogen is pyrolysed into hydrocarbons, producing fixed carbon. In the preheating zone, the oil shale is preheated by the residual heat from the exhaust gases. However, the reactions in the bed are complex, as each zone gradually undergoes a transition along the fixed bed, and the self-pyrolysis propagation is the result of a coupling between oil shale pyrolysis, oxidation and the transfer of heat and gas. Heat conduction and convection are the main modes of heat transfer[33].

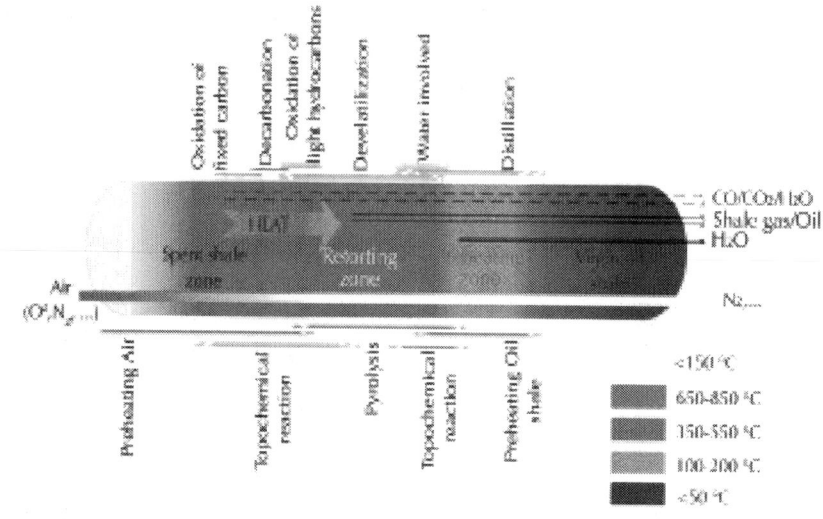

**Figure 5:** Description of the main phenomena occurring during the propagation of the self-pyrolysis front.

The topochemical reaction runs through the entire process, especially in the preheating and spent shale zones. In the preheating zone (Figure 5), oil shale is preheated to less than 200°C, producing water and carbon dioxide[23]. The water-filled pores in oil shale then form an autocatalytic system at 200–300°C, reducing the energy needed for chemical bond breakage and accelerating the conversion of kerogen into bitumen and into oil/gas. In the spent shale zone, with adequate oxygen, abundant heat is released through the complete oxidation of light hydrocarbons and the temperature increases rapidly (Figures 3a, b), accompanied by incomplete oxidation in areas with low concentrations of oxygen. Further oxidation of the fixed carbon consumes a large amount of oxygen, resulting in the temperature increasing to 800°C or higher[3]. This series of reactions may play a key role in the self-pyrolysis reaction process, which can be confirmed in time-interval trials. In the third trial, the temperatures of T1 and T2 increased to 500–700°C even after the oil shale completely pyrolysed in the previous two trials. Moreover, as evident in Figure 3d, when the sensor reached the highest temperature (~800°C), its right-side sensor was approximately 500°C; thus, according to the TG/DTG analysis in Figure 2, pyrolysis of the oil shale near the right-side sensor occurred. The small amount of fixed carbon (0.22%) in the residue also verified that almost all of the fixed carbon was consumed. All of these exothermic oxidations provided sufficient energy for further reactions and zone expansion. Moreover, the decomposition of kerogen, accompanied by changes in the microstructure of the solid phase, increased the porosity and opened channels for oil/gas drainage, which could improve gas transfer and heat conduction. In our contrast experiment, under an $N_2$ atmosphere, the supply of external heat was required through the whole pyrolysis of oil shale (~500°C); otherwise, the pyrolysis was terminated in a short time. The potential heat of the fixed carbon contained in the shale coke is not used, influencing the thermal efficiency of the process[34].

The topochemical reactions of oil shale can be easily triggered and repeated. After they have been triggered, oil shale can undergo pyrolysis spontaneously and smoothly with a limited amount of oxygen in the absence of external heat provision and the propagation speed of the pyrolysis front can be controlled by the air supply velocity. Compared with conventional oxygen-free pyrolysis, the heat produced by the topochemical reactions among the oxygen, water, fixed carbon and organics retained in the oil shale can not only accelerate the decomposition of kerogen but can also greatly reduce production costs. Moreover, the pyrolysis of oil shale triggered by topochemical heat is superior in its rapid heating rate and lower preheating gas requirement with a lower flow rate (Figure 3f). Hence, the low energy input and simplified process, together with the high extraction yield of the shale oil/gas, indicate that the proposed method is economically favourable towards retorting and in situ conversion.

Comparing the propagating fronts of the two experimental gas pipelines, we observed that, for the long gas pipeline, the hot gas flowed over the entire pipe until the end of the bed, at which point convective heat transfer with the surrounding shale occurred. As the temperature of T7 increased, the topochemical reaction started and heat was produced; the reaction fronts and gas flow acted in opposite directions. However, for approximately half of the experiment, the block of temperatures spread around T4 indicated that the gas was not only a reactant but also a heat carrier. The counter-current flow mode in the bed appears not to enhance the gas flow and the heat transfer; thus, the topochemical reaction gradually ceases. Therefore, horizontal drilling of a completely open channel may not necessarily be suitable for oil shale underground retorting. Hydraulic fracturing, which creates high-conductivity fissures with large areas in the oil shale stratum, appears to be more conducive to gas transmission and heat exchange.

The low-energy-consumption method described in this study provides an alternative route for further in situ industrial applications because of the possibilities of significant scale-up and lower environmental pollution. Notably, however, numerous critical issues must be considered before self-pyrolysis technology can be used for in situ industrial production. Firstly, further research should include a general performance evaluation and a comparative assessment of its energy use and cost. To ensure that the proposed self-pyrolysis is designed to maximise the topochemical reaction heat, a more detailed thermal analysis, including peak temperature and all of the incremental changes in the process parameters in all stages, should be conducted. For example, the peak temperature in our experiment was 800°C or higher (Figures 3b, d); however, the process could be optimised by limiting the temperatures to 600°C. The issue of how to provide sufficient heat through a complex in situ fracture network to trigger the reaction and drive the self-pyrolysis process also requires investigation. Moreover, the optimisation of oxygen content and gas flow for oil shale self-pyrolysis to improve the energy efficiency and reduce the costs is also necessary; such work is currently in progress.

# METHODS

## Materials

The oil shale used was obtained from the Gonglangtou mine located in Huadian (HD), China. The physical properties of the virgin oil shale and the spent oil shale from the self-pyrolysis experiments are summarised in Table 1.

## Experimental Apparatus

The experimental system shown in Figure 1a was specifically designed to enable the proposed self-pyrolysis experiments to be carried out. This system consisted of four main components: a gas supply and heating device, a fixed bed, a separation and recovery device, and a data monitoring and collection device. The gas supply and heating device comprised an air compressor or $N_2$ steel cylinder and a gas preheater. The fixed bed consisted of a horizontal cylindrical reactor chamber with an internal diameter of 325 mm and length of 2 m. The diameter was chosen to be sufficiently wide to limit heat losses from the walls and to facilitate treatment of the fuel gas. In addition to the gas channel, a groove was cut into the bottom of the bed to allow shale oil to discharge into a jerrican. The whole fixed bed can pack approximately 0.15 $m^3$ of oil shale. The fixed bed was constructed of a 5 mm thick stainless–steel material without a heating system and was surrounded by a 50 mm thick layer of refractory fibre. The separation and recovery device comprised a condenser, an oil and water separator, a jerrican and a water tank. The data monitoring and collection device comprised a flow meter, a temperature sensor, a pressure sensor and a paperless recorder.

## Gas Channel

Embedded gas channels were equipped into the fixed bed; these channels allowed the heated gas to enter into the bed. In this experiment, two circular gas pipelines with different lengths were used, as shown in Figures 1b, c. The short gas pipeline (Figure 1b) only reached the first sensor, whereas the long gas pipeline (Figure 1c) passed through the whole fixed bed, simulating the horizontal drilling mode. The external diameter of these two pipelines was 70 mm. The surface of both pipelines was full of long and narrow slits, which were beneficial to uniform gas discharge and could also prevent oil shale particles from falling into the pipeline. The short gas pipeline was generally used and is referred to as the default pipeline.

## Temperature Measurements

The fixed bed was finely instrumented, as shown in Figure 1. A group of seven in-line temperature sensors with a diameter of 8 mm and a length of 480 mm was equally spaced in the horizontal direction (from the left to the right of the reactor: T1, T2, T3, T4, T5, T6 and T7, as shown inFigures 1b, c), extending 200 mm into the reactor. At the length of 200 mm, four thermocouples (Figure 1d) were located at z =

0, 60, 120 and 180 mm (from the bottom to top of the sensor), enabling measurements of the temperature along the radial and axial cross-sections of the whole bed and monitoring of the change in temperatures and the progress of the front. Before the experiment, the temperature sensors were calibrated using a tubular quartz reactor; the range of temperature error was 2%, satisfying the precision requirements of our test.

## Self-pyrolysis Experiments

Firstly, a mass of large rubblized oil shale (2–100 mm, 110–120 kg), mimicking an anisotropic artificial fracture, was introduced into the fixed bed. For the self-pyrolysis experiments, hot air preheated by the gas preheater from ambient temperature to 400–500°C was fed into the bed via the gas pipeline. After the topochemical reaction was triggered, the heater was closed and the hot gas was replaced by ambient-temperature air. For the traditional oxygen-free pyrolysis process, pure hot $N_2$ was used as the only carrier gas. During the experiments, the gas flow was adjusted to control the reaction.

## Analytical Methods

TG and DSC were performed using a Netzsch STA449F3 (Germany) with a heating rate of 10°C $min^{-1}$ under both $N_2$ and air atmospheres (50 mL $min^{-1}$). The GC-MS analyses were carried out using an Agilent 6890/5973 N GC-MS instrument (America).

# ACKNOWLEDGEMENTS

This work was supported by the National Cooperative Innovation Project on Chinese Potential Oil and Gas Resources (Grant No. OSR-06), the Science and Technology Project of the Department of Jilin Province, China (Grant No. 20130302030SF), the Strategic Emerging Industry Development Projects of Jilin Province, China (Grant No. 2013Z050) and the National Natural Science Foundation of China (Young Scholars Program) (Grant No. 51404107). Thanks are due to Chuanbin Hou for assistance with the experiments and enlightening discussions.

# REFERENCES

1.  Na, J. G., Im, C. H., Chung, S. H. & Lee, K. B. Effect of oil shale retorting temperature on shale oil yield and properties. *Fuel* 95, 131–135, doi:10.1016/j.fuel.2011.11.029 (2012).

2.  Wang, S., Liu, J. X., Jiang, X. M., Han, X. X. & Tong, J. H. Effect of Heating Rate on Products Yield and Characteristics of Non-Condensable Gases and Shale Oil Obtained by Retorting Dachengzi Oil Shale. *Oil Shale* 30, 27–47, doi:10.3176/oil.2013.1.04 (2013).

3.  Martins, M. F., Salvador, S., Thovert, J. F. & Debenest, G. Co-current combustion of oil shale–Part 1: Characterization of the solid and gaseous products. *Fuel* 89, 144–151, doi:10.1016/j.fuel.2009.06.036 (2010).

4.  Rajeshwar, K., Nottenburg, R. & Dubow, J. Thermophysical properties of oil shales. *J. Mater. Sci.* 14, 2025–2052, doi:10.1007/BF00688409 (1979).

5.  Luik, H. *et al.* Upgrading of Estonian shale oil heavy residuum bituminous fraction by catalytic hydroconversion. *Fuel. Process. Technol.* 124, 115–122, doi:10.1016/j.fuproc.2014.02.018 (2014).

6.  Kelly, K. E., Wilkey, J. E., Spinti, J. P., Ring, T. A. & Pershing, D. W. Oxyfiring with CO2 capture to meet low-carbon fuel standards for unconventional fuels from Utah. *Int. J. Greenh. Gas Control* 22, 189–199, doi:10.1016/j.ijggc.2014.01.002 (2014).

7.  Taciuk, W. Does Oil Shale Have a Significant Future? *Oil Shale* 30, 1–5, doi:10.3176/oil.2013.1.01 (2013).

8.  Jaber, J. O. & Probert, S. D. Environmental-impact assessment for the proposed oil-shale integrated tri-generation plant. *Appl. Energ.* 62, 169–209, doi:10.1016/S0306-2619(99)00006-9 (1999).

9.  Han, X. X., Jiang, X. M. & Cui, Z. G. Studies of the effect of retorting factors on the yield of shale oil for a new comprehensive utilization technology of oil shale. *Appl. Energ.* 86,2381–2385, doi:10.1016/j.apenergy.2009.03.014 (2009).

10.  Speight, J. G. [Chapter 5 - In Situ Retorting] *Shale Oil Production Processes* [Speight, J. G. (ed.)] [123–138] (Gulf Professional Publishing, Boston, 2012).

11.  Brandt, A. R. Converting oil shale to liquid fuels: Energy inputs and greenhouse gas emissions of the Shell in situ conversion process. *Environ. Sci. Technol.* 42, 7489–7495, doi:10.1021/es800531f (2008).

12.  Abourriche, A. *et al.* Effect of toluene proportion on the yield and composition of oils obtained by supercritical extraction of Moroccan oil shale. *J. Supercrit. Fluid* 51, 24–28, doi:10.1016/j.supflu.2009.07.003 (2009).

13.  Deng, S. H. *et al.* Extracting hydrocarbons from Huadian oil shale by sub-critical water. *Fuel. Process. Technol.* 92, 1062–1067, doi:10.1016/j.fuproc.2011.01.001 (2011).

14.  Bolonkin, A., Friedlander, J., Neumann, S. & Strategic Solutions Technology, G. Innovative unconventional oil extraction technologies. *Fuel. Process. Technol.* 124, 228–242, doi:10.1016/j.fuproc.2014.01.024 (2014).

15.  Al-Makhadmeh, L., Maier, J., Al-Harahsheh, M. & Scheffknecht, G. Oxy-fuel technology: An experimental investigations into oil shale combustion under oxy-fuel conditions. *Fuel* 103, 421–429, doi:10.1016/j.fuel.2012.05.054 (2013).

16.  Qian, J. L. & Yin, L. *Oil Shale-Supplementary Energy of Petroleum.* (China Petrochemical Press, Beijing, 2011).

17.  Jiang, X. M., Han, X. X. & Cui, Z. G. Progress and recent utilization trends in combustion of Chinese oil shale. *Prog. Energ. Combust.* 33, 552–579, doi:10.1016/j.pecs.2006.06.002 (2007).

18.  Jiang, X. M., Han, X. X. & Cui, Z. G. New technology for the comprehensive utilization of Chinese oil shale resources. *Energy* 32, 772–777, doi:10.1016/j.energy.2006.05.001 (2007).

19.  Han, X., Kulaots, I., Jiang, X. & Suuberg, E. M. Review of oil shale semicoke and its combustion utilization. *Fuel* 126, 143–161, doi:10.1016/j.fuel.2014.02.045 (2014).

20.  Pan, Y., Zhang, X. M., Liu, S. H., Yang, S. C. & Ren, N. A Review on Technologies for Oil Shale Surface Retort. *J. Chem. Soc. Pak.* 34, 1331–1338 (2012).

21.  Gavrilova, O., Vilu, R. & Vallner, L. A life cycle environmental impact assessment of oil shale produced and consumed in

Estonia. *Resour. Conserv. Recy.* 55, 232–245, doi:10.1016/j.resconrec.2010.09.013 (2010).

22.  Niu, M. T., Wang, S., Han, X. X. & Jiang, X. M. Yield and characteristics of shale oil from the retorting of oil shale and fine oil-shale ash mixtures. *Appl. Energ.* 111, 234–239, doi:10.1016/j.apenergy.2013.04.089 (2013).

23.  Sun, Y. H. *et al.* Characterization of the oil shale products derived via topochemical reaction method. *Fuel* 115, 338–346, doi:10.1016/j.fuel.2013.07.029 (2014).

24.  Field, R. A., Soltis, J. & Murphy, S. Air quality concerns of unconventional oil and natural gas production. *Environ. Sci.-Process Impacts* 16, 954–969, doi: 10.1039/c4em00081a (2014).

25.  Strong, L. C. *et al.* Biodegradation in Waters from Hydraulic Fracturing: Chemistry, Microbiology, and Engineering. *J. Environ. Eng.-ASCE* 140, 8, doi: 10.1061/ (asce) ee.1943-7870.0000792 (2014).

26.  Hansom, J. & Lee, M.-K. Effects of hydrocarbon generation, basal heat flow and sediment compaction on overpressure development: a numerical study. *Petrol. Geosci.* 11, 353–360, doi: 10.1144/1354-079304-651 (2005).

27.  Martins, M. F., Salvador, S., Thovert, J. F. & Debenest, G. Co-current combustion of oil shale – Part 2: Structure of the combustion front. *Fuel* 89, 133–143, doi:10.1016/j.fuel.2009.06.040 (2010).

28.  Guo, H. F. *et al.* Retorting Oil Shale by a Self-Heating Route. *Energ. Fuel* 27, 2445–2451, doi: 10.1021/ef4000424 (2013).

29.  Akar, A. & Ekinci, E. Production of chemicals from oil shales. *Fuel* 74, 1113–1117, doi: 10.1016/0016-2361(95)00070-L (1995).

30.  Vossoughi, S., Willhite, G., El Shoubary, Y. & Bartlett, G. Study of the clay effect on crude oil combustion by thermogravimetry and differential scanning calorimetry. *J. Therm. Anal.* 27, 17–36, doi: 10.1007/BF01907318 (1983).

31.  Ali, M., Siddiqui, M. A. & Zaidi, S. M. Thermal analysis of crude oils and comparison with SIMDIST and TBP distillation data. *J. Therm. Anal. Calorim.* 51, 307–319, doi: 10.1007/BF02719032 (1998).

32.  Mothé, M. G., Carvalho, C. H., Sérvulo, E. F. & Mothé, C. G. Kinetic study of heavy crude oils by thermal analysis. *J. Therm.*

*Anal. Calorim.* 111, 663–668, doi: 10.1007/s10973-012-2574-1 (2013).

33. Shin, D. & Choi, S. The combustion of simulated waste particles in a fixed bed. *Combust. Flame* 121, 167–180, doi: 10.1016/S0010-2180(99)00124-8 (2000).

34. Gary, J. H. *An Assessment of Oil Shale Technologies.* (Office of Technology Assessment, Washington, D.C., 1980).

# Chapter 6

# The Oil Palm Wastes in Malaysia

N. Abdullah and F. Sulaiman

¹School of Physics, Universiti Sains Malaysia, 11800 Minden, Penang, Malaysia

## INTRODUCTION

Oil palm is the most important product from Malaysia that has helped to change the scenario of it's agriculture and economy. Lignocellulosic biomass which is produced from the oil palm industries include oil palm trunks (OPT), oil palm fronds (OPF), empty fruit bunches (EFB) and palm pressed fibres (PPF), palm shells and palm oil mill effluent palm (POME). However, the presence of these oil palm wastes has created a major disposal problem. The fundamental principles of waste management are to minimise and recycle the waste, recover the energy and finally dispose the waste. These principals apply to agro-industrial wastes such as palm oil residues as they do to municipal waste. We can simply no longer afford to dispose the residues when there is an economically useful alternative. We must first consider the current uses and disposal of mill residues in order to address the potential for recovery of energy in the palm oil industry. One of the unique aspects of Malaysian renewable energy sources is that the palm oil mill is self-sufficient in energy, using PPF, EFB and shell as fuel to generate steam in waste-fuel boilers for processing, and power-generation with steam turbines as described in Section 2.2.

World palm oil production in 1990 doubled to 11.0 million tonnes from 5.0 million tonnes in 1980, and by the year 2000, the production doubled to 21.8 million tonnes. Malaysia produced about half of the world palm oil production (10.8 million tonnes), thus, making Malaysia as world's largest producer and exporter of palm oil during this period [1]. In 2008, even though Malaysia had produced 17.7 million tonnes of palm oil based on 4,500,000 hectares of land used for its plantation, Indonesia became the world's largest producer and exporter of palm oil, replacing Malaysia as a chief producer [2,3] Palm oil has made impressive and sustained growth in the global market over the past four decades, and it is projected in the period 2016 – 2020, the average annual production of palm oil in Malaysia will reach 15.4 million tonnes [4]. In 1999, the land area under oil palm plantation is about 3.31 million hectares, and it has been projected that Sarawak will have about one million tonnes hectares of oil palm by the year 2010 [5].

The oil palm industry has always been linked to the environment because it is a land intensive industry. Any unplanned development will lead to the degradation of the forest systems, loss of habitats including plants and animals, extreme land degradation and pollution (water and airborne) due to the use of large quantities of pesticides and herbicides required to maintain the plantation. The Roundtable for Sustainable Palm Oil (RSPO) was established in recent years with the support from the government and Malaysia Palm Oil Council (MPOC). RSPO consists of palm oil producers, processors, traders, consumer goods manufacturers, retailers and non-governmental organizations (NGOs), and they will develop the principles and criteria of a sustainable palm oil industry, and facilitate the development of sustainable palm oil production. The proposed guidelines include commitment to transparency, compliance with all applicable local, national and ratified international regulations, adoption of sustainable cultivation practices (including water management, pesticide control and soil erosion), conservation of resources and biodiversity and community development [6]. The oil palm industry has long avoided the openings of virgin forest land, which thus minimize environment degradation and enhance the sustainability of oil palm growing. As initiatives, the Ministry of Plantation Industries and Commodities (MPIC) had announced in 2006, the RM20 million Malaysian Palm Oil Conservation Fund (MPOCF) with aims to help protect affected wildlife (including orang utan and other protected species) and to sustain biodiversity conservation programmes that are expected to be beneficial to both the industry and society.

Oil palm is the most important product of Malaysia that has helped to change the scenario of its agriculture and economy. Despite the obvious benefits, oil palm mill also significantly contributes to environmental degradation, both at the input and the output sides of its activities. On the input side, crude palm oil mills use large quantities of water and energy in the production processes, and on the output side, manufacturing processes generate large quantities of solid waste, wastewater and air pollution. The solid wastes may consist of empty fruit bunches (EFB), mesocarp fruit fibers (MF) and palm kernel shells (PKS). The liquid waste is generated from an extraction of palm oil of a wet process in a decanter. This liquid waste combined with the wastes from cooling water and sterilizer is called palm oil mill effluent (POME). During POME digestion, odor released into surrounding air, thus, reduces air quality in the surrounding lagoons area. Disposal of EFB into oil palm plantation without recovering remnant oil in the EFB contributes to oil spills. Incineration of EFB means wasting renewable energy source and heat which actually could be provided for boiler in palm oil mill. At present, PKS and MF wastes are used extensively as fuel for steam production in palm-oil mills. EFB is a resource which has huge potential to be used for power generation, currently not being utilized. The application of shells for road hardening has no impact to the environment, however, current practice is actually wasting potential renewable energy source. Methane gas is one among other greenhouse gases which can cause ozone depletion. However, at present, methane in biogas generates during POME digestion is not being utilized or captured and it just escapes into the atmosphere. Palm oil mill residues are currently underutilised; therefore, maximizing energy recovery from the wastes is desirable for both economic and environmental reasons.

All economic activity begins with physical materials and energy carriers such as fuels and electric power. Without materials, there might be no food and shelter technology; without energy, there might be no work, thus, no economic activity. The reliable sustainable resource is important to fulfill the need of energy. Oil palm waste is a reliable resource because of its availability, continuity and capacity for renewable energy solution. Furthermore, in current situation the presence of oil palm wastes has created a major disposal problem, thus, affect the environmental. The technological, economic, energy balance, and environmental considerations must be kept at a balance to meet the best solution of utilization oil palm wastes. There is abundance of raw materials available of the palm tree consisting of around 90% of biomass wastes and only around 10% of oil. About 90 million tonnes of oil palm fruit production was recorded in 1998; however, 43-45% of this was mill residues in the form of EFB, shell and fibre. Palm fronds and stems are currently underutilised, and the presence of these oil palm wastes has created a major disposal problem. Therefore, maximising energy recovery from the wastes is desirable for both the environmental and economic reasons. Direct combustion, gasification, pyrolysis, liquefaction, fermentation and anaerobic digestion are alternate conversion technologies available to maximise energy recovery. Therefore, sustainable development can be promoted by encouraging energy projects for the long term, utilising local skills and creating employment.

# OIL PALM INDUSTRY

Traditionally the oil palm (Elaeis guineensis) was grown in semi-wild groves in tropical Africa. It was first introduced to Malaysia for planting in the Botanical Gardens in Singapore in 1870 [7]. Germination takes around 3 months, after which the seedlings are planted in small plastic bags where they are left in a so-called pre-nursery for several months. They are transplanted into bigger plastic bags and grow in a nursery for several more months to a size of about 1 meter, before they are transplanted into a field at an age of around 1 year.

The new improved crosses begin to flower after less than one year of transplantation and produce their first bunches of fruit after less than 2 years. At this age, their leaves have a size of over 2 meters in height and diameter. During its young age, the trunk grows at a rate of about 35 to 75 cm per year and produces alternate rows of leaves, depending on its gene [8]. The base of the old leaves surround the stem and begin falling off at the age of 12 to 15 years [9]. By this time, growth and production have slowed down.

The number of leaves in an oil palm plant increase from 30 to 40 in a year at the age of 5 to 6 years. After that, the generation of leaves decreases to about 20 to 25 per year [9]. The average economic life-span of the oil palm is 25 years to 30 years [10]. A marked increase in the cultivation of oil palm began in 1960 [11], for which by the year 1990 onwards there was a peak in replanting. This provided a good opportunity to harness the by-products of the oil palm. During the re-plantation, the heights of the oil palm tree are in the range of 7 m to 13 m, with a width of between 45 cm to 65 cm, measuring 1.5 m from the surface of the soil. There are about 41 leaves in each frond of the mature oil palm tree. It is estimated that in the year 2000, the process of re-plantation would generate about 8.36 million tonnes dried biomass, consisting of 7.02 million tonnes of trunk and 1.34 million tonnes of leaves [5]. Due to the high moisture of about 70% fresh weight, the newly chopped tree trunk cannot be burnt in the plantation. To leave the old trunk for natural decomposition not only obstructs the re-plantation process but harbours insects that would harm the new trees as well. The tree trunk usually takes between five to six years to decompose [12].

Most crude palm oil mills harness the energy from the fibre and shell in their own low pressure boilers and normally, the EFB's are burnt causing air pollution or returned to the plantation. A 60 tonnes of fresh fruit bunches (FFB) per hour mill based within a 10,000 hectare plantation, can generate enough energy to be self sustaining and supply surplus electricity to the grid if it utilises all of its wastes. In order to provide a better understanding of the palm oil industry in Malaysia, the following sections give an overview of the oil palm industry in Malaysia including oil palm plantation and the mass balance of the oil palm industry as it is self-sufficient in energy.

## Malaysian Palm Oil Scenario

The first commercial oil palm estate in Malaysia was set up in 1917 at Tennamaran estate, Selangor. Palm oil is one of the seventeen major oils and fats in the world market. The government encouraged crop diversification from rubber to oil palm in the late 1950s. The area utilised for oil palm plantations in Malaysia has increased to 3.31 million hectares by the year 1999; where 62% of the total area is located in Peninsular Malaysia while Sabah and Sarawak 28% and 10%, respectively [4].

The oil palm fruit produces two distinct oils which are palm oil and palm kernel oil. Palm oil is obtained from the mesocarp while palm kernel oil is obtained from the seed or kernel. Palm oil is used mainly for the production of margarine and compounds in cooking fats and oils and also for the production of candles, detergents, soap and cosmetic products. Production of palm kernel oil is about 12% of the production of its palm oil.

The success of the Malaysian palm oil industry is the result of the ideal climatic conditions, efficient milling and refining technologies and facilities, research and development, and efficient and adequate management skills. Practically all palm oil mills generate their own heat and power through the co-generation system [13]. The Malaysian government is fully committed to the expansion of the industry and encourages global expansion of palm oil production. Palm oil is now readily accepted globally and Malaysia has exported palm oil to more than 140 countries in the world.

Most palm oil is currently produced in South East Asia, even though the oil palm is originally an African crop, which was introduced to South East Asia in the 19th century. The two largest producers are Malaysia and Indonesia, who together account for roughly 85% of the world palm oil production [14]. In 2004 Malaysian production exceeded Indonesian production. However, the US Department of Agriculture notes that mature palm area in Indonesia is being expanded from 5 to 8 million hectares, which should easily overtake Malaysia in the near future [15]. There are plans for expansion of palm area in South America [16] and Africa [17], both of which in principle offer large tracts of suitable tropical land. Compared to the potential expansion, however, these plans are embryonic and current production is low and largely for domestic consumption.

Palm oil and related products represented the second largest export of Malaysia in the first nine months of 2005, after electronics, but just ahead of crude oil [18]. In 2005, Malaysian palm oil production is projected to reach approximately 15 million tonnes (301,000 barrels per day), which is very close to the actual value of 14.96 million metric tons recorded by Malaysian Palm Oil Board (MPOB) [19,20]. By comparison, Malaysian petroleum production in 2004 is estimated at 43 million tonnes (855,000 barrels per day), of which 16 million tonnes (321,000 barrels per day) were exported. Domestic petroleum demand of 26 million tonnes represented 44% of the total energy demand of 60 million tonnes of oil equivalent [21].

The total oil palm planted area in Malaysia increased by 2.8% to 4.17 million hectares in 2006. The area expansion occurred mainly in Sabah and Sarawak with a combined growth of 4.5% compared to 1.6% in Peninsular Malaysia [22]. Sabah remained the largest oil palm planted state with 1.24 million hectares or 30% of the total planted area. Table 2.2 shows the oil palm planted areas by state in Malaysia for 2005 until 2008 (in hectares) [22, 23].

**Table 1:** Oil Palm Planted Area 2005 - 2008 (Hectares) [22, 23]

| State | 2005 | 2006 | 2007 | 2008 |
|---|---|---|---|---|
| Johor | 667,872 | 671,425 | 670,641 | na |
| Kedah | 75,472 | 76,329 | 75,096 | na |
| Kelantan | 89,886 | 94,542 | 99,763 | na |
| Melaka | 52,015 | 52,232 | 49,113 | na |
| N.Sembilan | 155,164 | 161,072 | 170,843 | na |
| Pahang | 606.821 | 623,290 | 641,452 | na |
| Perak | 340,959 | 348,000 | 350,983 | na |
| Perlis | 278 | 258 | 260 | na |
| P.Pinang | 14,074 | 14,119 | 13,304 | na |
| Selangor | 132,100 | 128,915 | 129,315 | na |
| Terengganu | 163,967 | 164,065 | 161,287 | na |
| **Peninsular Malaysia** | **2,298,608** | **2,334,247** | **2,362,057** | - |
| Sabah | 1,209,368 | 1,239,497 | 1,278,244 | na |
| Sarawak | 543,398 | 591,471 | 664,612 | na |
| **Sabah & Sarawak** | **1,752,766** | **1,830,968** | **1,942,856** | - |
| **Malaysia** | **4,051,374** | **4,165,215** | **4,304,913** | **4,487,957** |

The production of crude palm oil increased by a further 6.1% to 15.9 million tonnes in 2006 from 15.0 million tonnes the previous year as shown in Figure 2.3. Figure 2.4 shows that the increase was mainly attributed to the expansion in matured areas by 2.0% and rise in the average fresh fruit bunches yield per hectare by 3.8% to 19.6 tonnes due to better management and agricultural inputs.Figure 2.4 also shows that the oil yield per hectare had increased by 3.4% to 3.9 tonnes, despite the oil extraction rate (OER) declining marginally by 0.5% to 20.04% as shown in Figure 2.5. The decrease in OER in the years 1993 to 2001 which is significant is due to the global recession accounting for a lower demand of export market. However, despite a weak global economy, there is a significant recovery in 2002 as the government implemented prudent policies to assist the Malaysian oil palm industry. These include the expansion of oil palm in matured areas and the campaign on improved productivity in the oil palm industry, coupled with providing competitive prices of oil palm, liberalization of export duties and the encouragement of counter-trades for higher exports [24]. Crude palm kernel oil production rose by 6.1% to 1.96 million tonnes in tandem with a 4.1% growth in palm kernel production as shown in Figure 2.3 [18, 23].

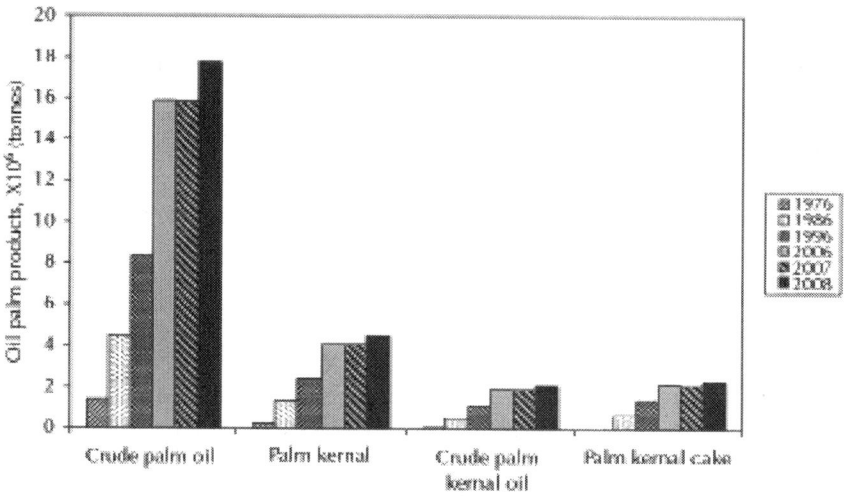

**Figure 1:** Production of Palm Oil Products every Ten Years from 1976 – 2006, 2007 and 2008 [18, 23].

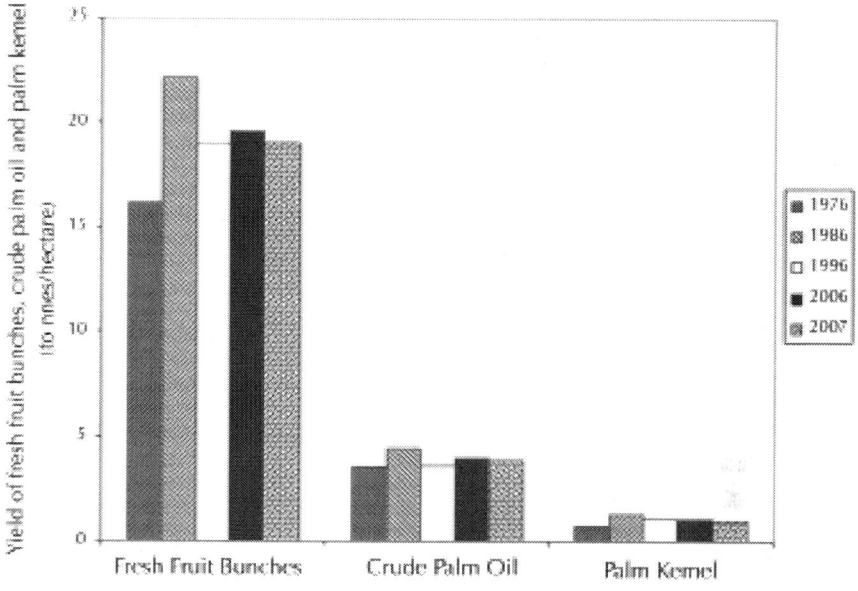

**Figure 2:** Yield of Fresh Fruit Bunches, Crude Palm Oil and Palm Kernel every Ten Years from 1976 – 2006 and 2007 [18].

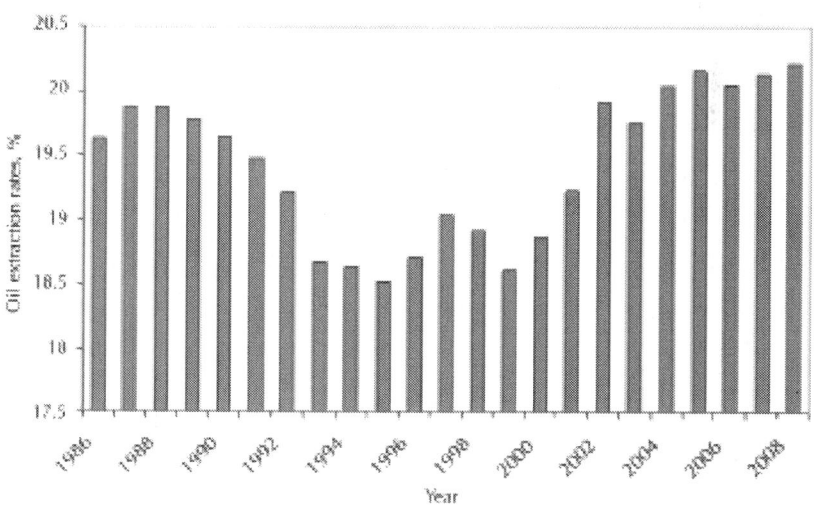

**Figure 3:** Annual Oil Extraction Rate (OER) for 1986-2008 (%) [22, 23].

The rapid expansion of oil palm cultivation has raised concerns about the sustainability and environmental impact of oil palm plantations, in particular with regard to biodiversity, destruction of old growth rainforest and air pollution [25,26]. To illustrate the potential impact, it is worthy to reflect on the fact that with a palm oil yield of 4 tonnes per hectare tropical forest of roughly the size of the United States would be required to satisfy current world crude oil demand. Increased yields are one avenue for reducing the area imprint for oil palm plantations. It is estimated, based on fundamental factors and actual yields achieved on experimental plots that yields as high as approximately 10 tonnes per hectare may eventually be achievable [27]. At these yields, current world oil demand could be met on roughly 4 million square kilometres, which is 40% of the area of the United States, or over half the land mass of Brazil.

# The Mass Balance of the Oil Palm Industry

The palm oil mill is self-sufficient in energy, using waste fibre and shell as fuel to generate steam in waste-fuel boilers for processing, and power-generation with steam turbines. As an example, The Federal Land Development Authority (FELDA) palm oil mill in Sungai Tengi, Selangor, Malaysia, employs the standard oil extracting process [28]. In the standard milling process, used in the factories with a milling capacity of over 10 tonnes of raw material per hour, water is added into a digester [29]. More than 19.7 million tonnes FFB were processed in 2000 [28]. The standard sized mills processing 60 tonnes/hour of fruit bunches normally produce 40 tonnes/hour of steam. Part of the steam is used to generate 800 kW of electricity and the rest is used as process steam. It is estimated that the total generating capacity of the mills is about 200 MW [28]. Typically palm oil mills use fibre and shell as a boiler fuel to produce process steam for sterilisation, etc and also possibly for electricity generation to supply electricity for other parts of the mill complex. These oil palm wastes make oil palm mills self-sustainable in energy. The shell and fibre alone can supply more than enough energy to meet the mill's requirements using low pressure relatively inefficient boilers. The EFB have traditionally been burnt in simple incinerators, as a means of disposal and the ash recycled onto the plantation as fertiliser. However, this process causes air pollution and has now been banned in Malaysia, furthermore, under this route of disposal, no energy is recovered. Alternatively EFB can be composted and returned to the plantation, or returned directly as mulch. Figure 2.1 shows a proposed plan for the operational process and product of the palm oil industry if EFB is used as fuel beside palm shell and fibre.

Referring to Figure 2.1, as the fresh fruit bunches reach the processing plant, the sterilisation process begins with the steam temperature at 140°C, pressure at 2.5 to 3.2 kg/cm$^2$ for 50 minutes [28]. After this process, the stripping process will take over. In the stripping process, a rotating divesting machine is used to separate the sterilized oil palm fruit from the sterilized bunch stalks. The empty fruit bunches (EFB) will fall in the collector and are brought to the burning place as a fuel. After the bunches have been stripped, the sterilised fruits are fed into a digester where water at 80°C is added. This is performed in steam-heated vessels with stirring arms, known as digesters or kettles. The most usual method of extracting oil from the digested palm fruit is by pressing. The type of press used in this palm oil is the screw type press.

The crude oil extracted from the digested palm fruit by pressing contains varying amounts of water, together with impurities consisting of vegetable matter, some of which is dissolved in the water. Centrifugal and vacuum driers are used to further purify the oil before pumping it into a storage tank. When the digested fruit is pressed to extract the oil, a cake made up of nuts and fibre is produced. The composition of this cake varies considerably, being dependent on the type of fruit. The cake is given a preliminary breaking treatment before being fed into the nut/fibre separator called depericarper. When the fibre has been separated from the nuts, the latter can then be prepared for cracking. Any uncracked nuts must be removed and recycled and the shell separated from the kernels. The waste fibre and shell are also transported to the burning place as a fuel. The kernels are packed and sold to kernel oil mills.

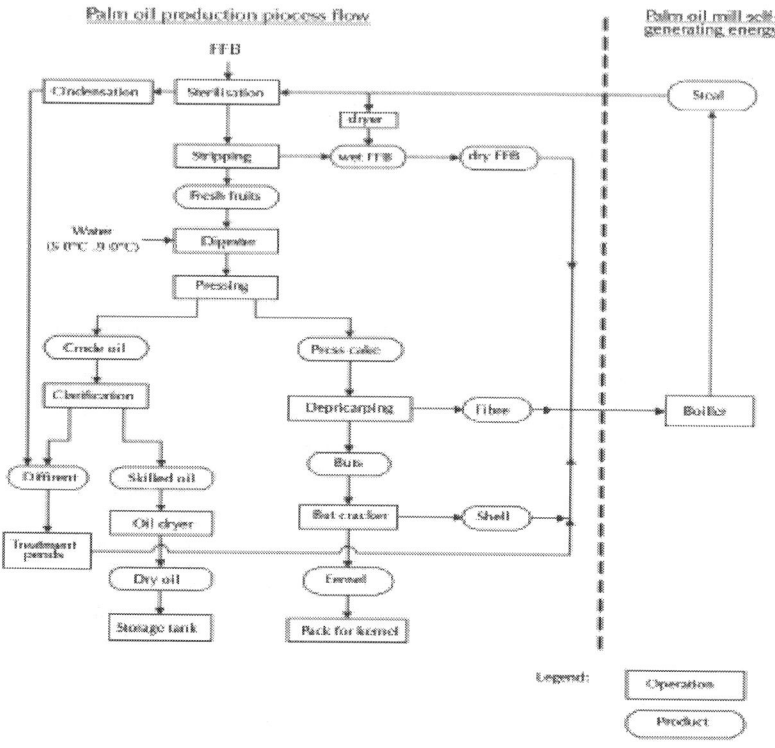

**Figure 4:** Proposed plan for operation of a Palm Oil Mill (adapted from [28]).

Palm oil mills in Malaysia typically meet most of their electricity and process steam requirements by burning some of the wastes, with energy for start-up generally being provided by back-up diesel [13, 28,30]. Not all of the wastes are burnt. For each kg of palm oil, electricity consumption is around 0.075-0.1 kWh and steam demand around 2.5 kg. This represents a steam to electricity ratio of around 20 to 1 and could be met by burning 0.3-0.4 kg of waste. As the boiler efficiency is only around 70%, actual consumption is correspondingly higher [31]. Little effort was made in the past to optimise process steam consumption or boiler or turbine efficiency, as the fuel was substantially treated as a waste that was incinerated to be disposed of. The electricity co-generated in Malaysian palm oil mills therefore only amounts to roughly 1-1.5 billion kWh or less than 2% of 2003 generation of over 82 billion kWh. To illustrate the kinds of waste available, the process flow of a palm oil mill is summarised in Figure 2.2 (simplified from

[28]) and a typical product stream distribution is shown in Table 2.1 (adapted from [30]). The total product stream distribution in oil palm mills is greater than 100% in wet basis as extra water is added during the process, for example during sterilization with steam. Most of this water ends up in POME.

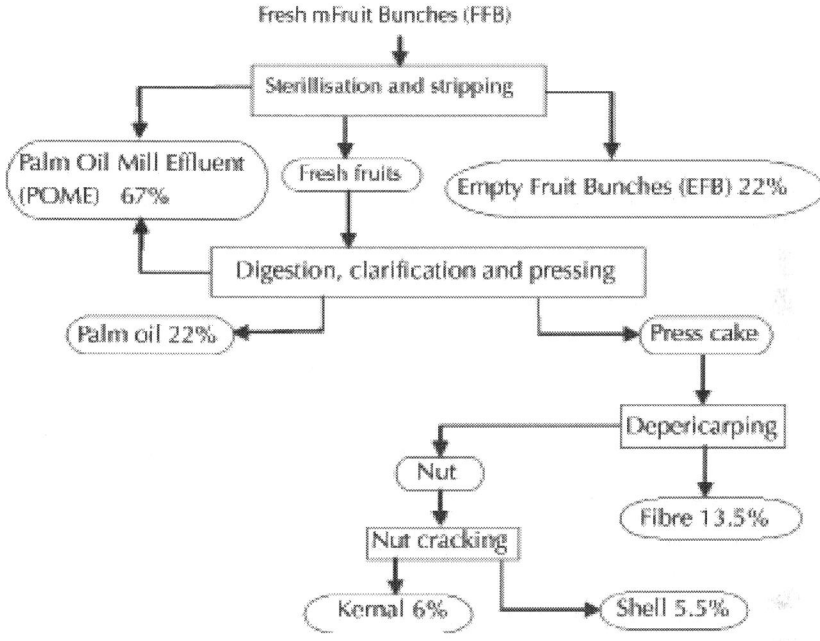

**Figure 5:** Simplified process flow diagram of an oil palm mill.

Note: main waste streams in bold, all percentages on wet FFB basis.

As can be seen in Table 2.1, the moisture content of fresh EFB is very high. Typically it is over 60% on a wet EFB basis. Consequently, it is a poor fuel without drying and presents considerable emissions problem that its burning is discouraged by the Malaysian government. Palm oil mills therefore typically use shell and the drier part of the fibre product stream, rather than EFB, to fuel their boilers [31]. Palm Oil Mill Effluent (POME) is so wet that it is usually treated by anaerobic digestion before the discharge of the effluents [32].

For each kg of palm oil, roughly a kg of wet EFB is produced. As over 60% of the wet EFB consists of water, and the heating value of the dry EFB is roughly half that of palm oil, the energy obtainable from the EFB product stream amounts to roughly 0.2 kg of oil equivalent per kg of palm oil. Based on Malaysia's 2005 palm oil production of 15 million tonnes, the energy value of the EFB waste is therefore around 3 million tonnes of oil equivalent, which would amount to $1.2 billion for an assumed $400 per tonne ($55 per barrel).

**Table 2:** Typical product stream distribution in oil palm mills [30]

|  | Wet FFB basis | | Dry FFB basis | |
|---|---|---|---|---|
|  | (tonnes per hectare) | % FFB | (tonnes per hectare) | % FFB |
| FFB | 20.08 | 100 | 10.6 | 100 |
| Palm oil | 4.42 | 22.0 | 4.42 | 41.7 |
| Palm kernel | 1.20 | 6.0 | 1.20 | 11.4 |
| EFB | 4.42 | 22.0 | 1.55 | 14.6 |
| POME | 13.45 | 67.0 | 0.67 | 6.3 |
| Shell | 1.10 | 5.5 | 1.10 | 10.4 |
| Fibre | 2.71 | 13.5 | 1.63 | 15.4 |
| Total | 27.3 | 136.0 | 10.6 | 99.8 |

As mentioned before, most crude palm oil mills harness the energy from the fibre and shell in steam boilers. However, the introduction of advanced cogeneration (combined heat and power) also can play a role in combatting climate change, as well as introducing significant economic benefits. Through cogeneration, the costs of energy will be cut because it uses fuels at high conversion efficiencies can reduce the emissions of carbon dioxide and other pollutants. However, it is only worth doing if one can sell the additional surplus energy (electricity) to customers at an economical rate. Today, the ability to sell electricity into the local grid provides an opportunity to turn waste into a valuable commodity.

## Options for the Disposal of Oil Palm Wastes

The total land area in Malaysia amounts to 32.90 million hectares. According to Hoi and Koh [32], the major agricultural crops grown in Malaysia are rubber (39.67 %), oil palm (34.56 %), rice (12.68 %), cocoa (6.75 %) and coconut (6.34 %) which indicated that major production of the agricultural sector had been rubber derived products including wood residues, however, by 1995 oil palm products became more significant [34].

Lignocellulosic biomass which is produced from the oil palm industries include oil palm trunks (OPT), empty fruit bunches (EFB), fronds, palm pressed fibres (PPF) and shells. Table 2.3 shows the breakdown of wastes from palm oil production in 2007 [35].

**Table 3:** Wastes from palm oil production [35]

| Wastes | **Quantity (ktonnes)** |
|---|---|
| Fronds | 46,837 |
| Empty fruit bunches (EFB) | 18,022 |
| Palm pressed fibres (PPF) | 11,059 |
| Oil palm trunks (OPT) | 10,827 |
| Shell | 4,506 |

One of the major characteristics of the forestry and agricultural sector is the production of large quantities of processing residues that have no economic value other than energy generation. Their presence in recent years has created a major disposal problem due to the fact that open burning is being discouraged by the Department of Environment in Malaysia. Other than biomass from the plantations, the palm oil industry also produces other types of waste in large quantities mainly EFB, PPF, shell and palm oil mill effluent (POME). Table 2.4 shows the breakdown of product or waste from each bunch of fresh fruit (FFB) [36].

**Table 4:** Products/wastes from each bunch of FFB [36]

| Products/Wastes | Percentage by weight to FFB (dry basis) |
|---|---|
| Palm oil | 21 |
| Palm kernel | 7 |
| Fibre | 15 |
| Shell | 6 |
| Empty fruit bunches | 23 |

| POME | 28 |
|---|---|
| Total | 100 |

The EFB are usually air dried until the moisture content reaches about 40% when it is ready to be used as fuel in the palm oil processing plant [37]. The burnt waste is then used as fertiliser in plantations [38]. Other than that, EFB were also used in the plantations as a mulch, thus, can reduce the applied fertiliser cost and is a step towards environmental conservation by reducing dependence on fossil fuel required for the manufacture of inorganic fertilizer [39]. It is claimed that using the EFB as mulch has several advantages for the nutritional sustainability of the plantation. Some plantation owners claimed that the benefits of EFB as a fertiliser and as a soil conditioning agent are significant, because it releases nutrients slowly to the soil via microorganisms therefore effectively recycling the plant nutrients. It improves the soil structure due to better aeration, increases the water holding capacity and increases the soil pH, whilst other mill owners welcomed alternative methods of disposal. This is due to the inconvenience of handling and transporting, as well as the costs and problems concerning disposal of the waste on the plantation. However, open burning is no longer allowed by the authority because this process causes air pollution and by this means of disposal no energy is recovered [40].

Oil-palm fronds have been successfully used as a substitute for tropical grasses by ruminant producers in Malaysia [41]. Nowadays, the PPF is usually burnt in the palm oil processing plant as fuel and the excess is disposed of in the plantations [42]. The PPF are burnt in a boiler with some palm shells to produce the power for running the mill (self-sufficient). The boilers used are normally of grate-type beds which are manufactured locally [13]. Most of the crude palm oil mills harness the energy from the shell and fibre in their own low-pressure boilers and normally the oil palm trunk would be left to decompose naturally at the plantation [37]. This practice not only disturbs the process of plantation due to the low decomposition rate, it also encourages the spread of diseases and insects like *rhinoces beetles* and *ganoderma* that are harmful to the plantation [37]. Moreover, most of the plantations have to adopt the push-felling technique and trunk-shredding which leads to burning [43].

The utilization and generation of oil palm biomass is widely accepted and offers benefits for rural areas related to employment, rural infrastructure, the conservation of cultivated areas and hence the attractiveness of rural regions. The new markets for Malaysia can be developed, especially for developing countries, where oil palm biomass has a higher contribution to the overall energy supply. Also the establishment of an industry related to 'oil palm biomass for energy" technology could be supported.

# UTILIZATION OF OIL PALM WASTES

Another route to obtain more energy from oil palm plantations is the more efficient use of oil palm biomass other than the palm oil. There are no detailed statistics for oil palm dry matter production. Such statistics are only compiled for palm oil, palm kernel and fresh fruit bunches (FFB). Rough extrapolations, however, can be made based on estimates of the ratio of palm oil to other dry matter. For each kg of palm oil roughly another 4 kg of dry biomass are produced; approximately a third of which is found in FFB derived wastes and the other two thirds is represented by trunk and frond material [27,30,44]. On an energy basis, the palm oil represents roughly a third of the biomass yield, as it has roughly twice the heating value of the other oil palm dry matter, which therefore amounts to approximately 2 kg on a palm oil equivalent basis. Based on 2005 production, around 30 million metric tonnes of oil equivalent of non-palm oil dry biomass matter were available for energy production from Malaysian palm oil plantations, or in other words approximately half of 2004 total primary energy demand. Only a small fraction of this potential was used, and that vary inefficiently. Open burning is still too common and responsible for substantial air pollution problems in South East Asia, indicating that other solutions urgently need to be found. Some of the biomass is used for mulching and as fertiliser, though this use is limited by labour and logistical limitations and concerns about encouraging oil palm pests [45].

Generally, oil palm mills generate a numbers of oil palm wastes. The oil palm wastes contribute about RM6379 million of energy annually [46]. However, there is much to be done to optimise the utilization of oil palm wastes for cogeneration in Malaysia. Various studies conducted in Malaysia have indicated that the used of biomass as a source of energy is one of the most promising ways of effectively using the residues. Some of the commercial projects and research activities are include treatment of palm oil mill effluent [47,48], pyrolysis of oil palm shell [49], chars from oil palm waste [50], solid biofuels from biowastes [51], briquetting of palm fibre and shell [36], palm oil effluent as a source of bioenergy [52] ethanol fermentation from oil palm trunk [53] and converting oil palm trunks and cocoa wood to liquid fuels [37]. In the following sections, potential uses of oil palm wastes are presented.

## Potential Uses of Pome

POME is the effluent from the final stages of palm oil production in the mill. It is a colloidal suspension containing 95-96% water, 0.6-0.7% oil and 4-5% total solids including 2-4% suspended solids [54]. Most palm oil mills and refineries have their own treatment systems for POME, which due to its high organic content is easily amenable to biodegradation. The treatment system usually consists of anaerobic and aerobic ponds. However, because of silting and short circuiting many do not reach discharge standards to water courses. This situation can be significantly improved by introducing enclosed anaerobic digestion systems which reduce the biological oxygen demand (BOD) of the effluent and capture methane, one of the more potent greenhouse gases. The energy in the methane can then be recovered, either as a supplementary boiler fuel, or in a biogas engine generator. For each tonne of crude palm oil (CPO) produced, about an average of 0.9-1.5m$^3$ POME is generated. The biological oxygen demand (BOD), chemical oxygen demand, oil and grease, total solids and suspended solids of POME ranges from 25000 to 35000 mg/L, 53630 mg/L, 8370 mg/L, 43635 mg/L and 19020 mg/L respectively [55]. Therefore, this had created environmental problem because the palm oil mill industry in Malaysia produces the largest pollution load into the rivers throughout the country [56]. However, POME contains high concentrations of protein, nitrogenous compounds, carbohydrate, lipids and minerals that could be converted into useful material using microbial process [57, 58]. As example, bio-gas can be produced by processing POME through anaerobic treating system. Anaerobic digestion is a series of processes in which microorganism break down biodegradable material in the absence of oxygen. About 400m$^3$ of bio-gas produced from 100 tonnes of POME, of which this amount of POME had been released during processing of 20 tonnes of fresh fruit bunches [59-61].

Currently, fertilizers is also derived from POME and used in the farms and vegetation areas [62]. It is also found that the gas composition contained hydrogen (66-68%) and carbon dioxide (32-34%) that can be produced from POME using anaerobic micro flora and this generated gas is free from methane [63]. At present, a renewable energy power plant developer in Malaysia, known as Bumibiopower is in the progress of setting up a plant from methane extraction and power generation using POME near Pantai Remis at the west coast of Peninsular Malaysia. A closed anaerobic system is installed to produce and collect consistently high quality of methane-rich biogas from POME. The installation of a generator of size between 1 and 1.5 MW is also included in this project [64].

# Potential Uses of Bio Oil Derived From Oil Palm Wastes

Bio-oil is a renewable, which is produced from biomass through a process known as fast pyrolysis. Fast pyrolysis represents a potential route to upgrade the biomass to value added fuels and renewable chemicals. There is an urgent need to develop a sustainable energy supply as the impact of burning fossil fuels on our climate is becoming ever more obvious and the availability of fossil fuels is decreasing. Bio-oil contributes to the reduction of greenhouse gas emissions and it offers several advantages, as it is easy to use, to store, and to transport. Bio-oil that can be extracted from dried biomass including dried oil palm wastes is currently under investigation as a substitute for petroleum [65]. Bio-oil contains fragments of cellulose, hemicelluloses, lignin, and extractives and they are typically brown liquids with a pungent odor. For woody feed stocks, temperatures around 500°C together with short vapour residence times are used to obtain bio-oil yields of around 70%, and char and gas yields of around 15% each [66]. Bio-oil is a high density oxygenated liquid, which can be burned in diesel engines, turbines or boilers, though further work is still required to demonstrate long term reliability [67]. It is also used for the production of speciality chemicals, currently mainly flavourings. Renewable resins and slow release fertilisers are other potential applications, which have been the subject of research [68]. At this stage, fast pyrolysis is a novel and relatively untested technology. There are several pilot plants in

North America and Europe, but no consistent track record yet outside of the manufacture of flavourings.

To date, fast pyrolysis of biomass has received very limited attention by researchers in Malaysia. Normally, fibre and shells are burnt in the palm oil processing plants to generate fuel to produce power for running the mill (self-sufficient) [13,69]. So far, research involving fast pyrolysis has been carried out by Universiti Teknologi Malaysia and Universiti Malaya on oil palm shell, rubber waste and rice husk waste, scrapped tyres and tubes [70-74]. One of the authors of this book and the research group from MPOB investigated on the fast pyrolysis of empty fruit bunches (EFB) [75,76].

The utilisation of bio-oil derived from pyrolysis process of oil palm wastes to substitute for synthetic phenol and formaldehyde in phenol formaldehyde resins is possible. Phenol can be used to manufacture moulding products for automotive parts, household appliances, and electrical components; in bonding and adhesive resins for laminating, plywood, protective coating, insulation materials, abrasive coating; in foundry industries for sand moulds and cores. However, producing resins from bio-oil has received very limited attention by researchers in Malaysia and still in research stage. A group of researchers from Universiti Teknologi Malaysia had studied the extraction of phenol from oil palm shell bio-oil [77]. They found that the quantity of phenol in the extracted oil was 24.2 wt% of total extracted oil.

In 2005, with the co-operation between Malaysian based Genting Sanyen Bhd and BTG Biomass Technology Group BV from The Netherlands, the first commercial bio-oil plant has already started production in Malaysia on a scale of 2 t/hr [78,79]. The main achievements of this project are more than 1,000 tonnes of bio-oil have been produced, the bio-oil is co-fired, replacing conventional diesel in a waste disposal system located 300 km from site, maximum capacity of the plant so far is about 1.7 t/hr on a daily continuous basis, the bio-oil quality can be controlled by the operating conditions, the drying of EFB to 5 wt.% moisture is possible using the excess heat from the pyrolysis process, the energy recovered from the process can be used effectively for drying the wet EFB, and potentially to generate the electricity required. Indeed, this is a breakthrough step in Malaysia for the utilisation of oil palm wastes as a source of bio-oil [80].

# Potential Uses of Dry Residues from Oil Palm Wastes

The main products produced by the palm oil mills are crude palm oil and palm kernels. However, it also produces huge quantities of residues such as fibre, shell and empty fruit bunches as shown inFigure 2.2. Dry residues from oil palm wastes can be utilised to produce various types of products. EFB had been studied to convert into paper-making pulp by the researches from MPOB because EFB can be categorized as fibrous crop residues know as lignocellulosic residues. The high number of fibres/unit weight indicates the paper from EFB would have good printing properties and a good formation within paper making. EFB could produce thin, high quality printing paper, speciality papers for example for cigarette and photographic papers and security papers. The total chlorine-free methods had been used to bleach the pulp for producing paper [59,81]. Products such as paper and pulp that are obtained by processing the oil palm wastes can be used in many ways such as cigarette paper and bond papers for writing [82]. Normally, the excess shell are used to cover the surface of the roads in the plantation area.

Various types of wood such as saw-wood and ply-wood or lumber had been produced from oil palm trunk. Oil palm trunks have been chipped and waxed with resin to produce pre formed desk tops and chair seats for schools. The furniture is characterised for resistance against knocks, scratches, ink, termites and fungus The ply-wood or lumber can be utilised as core in producing blackboard. The saw-wood is used for furniture but it is not suitable as building material due to its low specific density. It was found that the strength of the ply-wood made from oil palm trunk was comparable with the commercial ply-wood. The particle board with chemical binders also can be produced from oil palm trunk. Some of the oil palm trunks are mixed with EFB and palm fibres to be combusted to produce energy [81,83,84]. Besides this, the palm shell and palm fibres have been convert of into briquettes in a study [36].

Medium density fibre-boards and blackboards can be produced from EFB and palm fibre [84,85]. Currently, the MDF industry has 14 plants with a total annual installed capacity of 2.9 million. The total export of MDF was RM1.2 billion in 2008. The industry has started utilising *acacia mangium*and mixed hardwood to produce MDF as alternatives to rubber wood. At present, Malaysia is the world's third largest exporter of MDF, after Germany and France. MDF from Malaysia has attained international standards such as British (BS), European (EN), Asia-Pacific: Japan Australia and New Zealand (JANS) standards [86]. High-density fiberboard (HDF), also called hardboard, is a type of fiberboard, which is an engineered wood product. It is similar to MDF, but is denser and much harder and stronger because it is made out of exploded wood fibers that have been highly compressed. Agro-Bio Fibre Sdn Bhd in Malaysia holds the patent for the EFB-based MDF over the last 10 years, has invested RM30 million to develop the technology to produce MDF and other products from the oil palm wastes. This company had signed a MoU with the Forest Research Institute of Malaysia (FRIM) to develop HDF used mainly for the production of floorboards that would use 100% EFB as its raw material [87].

Oil palm fibre is non-hazardous biodegradable material extracted from empty fruit bunch that are considered as waste after the extraction oil palm fruits. The fibres are clean, non-carcinogenic, and free from pesticides and soft parenchyma cells. Palm fibres are versatile and stable and can be processed into various dimensional grades to suit specific applications such as erosion control, mattress cushion production, soil stabilization, horticulture and landscaping, ceramic and brick manufacturing, paper production, acoustics control, livestock care, compost, fertilizer and animal feed. Palm fibres can also be used as fillers in thermoplastics and thermoset composites which have wide applications in furniture and automobile components. Production of thermoplastic and thermostat composites has reached commercialization stage when PROTON (Malaysian national car maker) entered into agreement with PORIM (Palm Oil Research Institute of Malaysia) [88,89].

Similar to EFB, according to a study fronds from oil palm trees can be converted into pulp [90]. Oil palm fronds also can be processed as roughage source for ruminants such as cattle and goats [91]. A new product known as oil palm frond based ruminant pellet can be used as balanced diet for fattening beef cattle which is developed by the Malaysian Agricultural Research and Development Institute (MARDI) [91].

Oil palm ash (OPA) can be utilised as an absorbent for removing pollutant gases such as nitrogen oxide and sulphur oxide. The combustion of oil palm fibre and shell as boiler fuel to generate steam in palm oil mill will produce OPA. It was found that OPA contains high amount of calcium, silica, potassium and alumina which can be utilised to synthesize active compounds to absorb the pollutant gases into absorbent [92,93]. The presence of some functional groups such as hydroxyl, lactone and carboxylic in oil palm shell have a high affinity towards metal ions. Thus, the charcoal derived from oil palm shell can be coated with chitosan to use as a remover of heavy metal especially chromium from wastewater industry; however, it is still at research stage [94].

Processing the oil palm wastes such as EFB, fibre, shell and palm kernel cake into a uniform and solid fuel through briquetting process will be an attractive option. Palm kernel cake is a by-product of crushing and expelling oil from palm kernel. Briquetting is a process of compacting loose material to form a homogeneous and densified product. The material can be densified into briquettes at high temperature and pressure using screw of extrusion techniques either with or without binder addition. Oil palm briquettes are often favoured for household and industrial heating unit operation such as boiler because of their enhanced physical properties, as well as being easy to handle and feed. According to a study, the equilibrium moisture content for the briquettes made of palm fibre and palm shell is about 12 mf wt.% [36]. It was found that briquettes made from 100% pulverised EFB exhibited good burning properties. It is recommended to blend with sawdust in order to produce better quality briquettes from EFB and palm kernel cake [95]. Oil palm briquettes can be used as fuel in producing steam, district heating and electricity generation for larger commercial scale. The local sawdust briquettes or charcoal briquettes are rarely used in the local market because it could not compete with the availability of cheap fuels such as charcoal and wood which are widely used in the rural areas and restaurants [96]. Therefore, the products are exported for oversea markets [97].

One of the promising technologies which utilise the oil palm wastes or plant matter involves the production of carbon molecular sieve (CMS) from lignocellulosic materials. Production of CMS from oil palm wastes which are cheap and abundant carbon source will enhance the economical feasibility of adsorption process. A CMS is a material containing tiny pores of a precise and uniform size that is used as an adsorbent for gases and liquids, and normally it is used to separate nitrogen from the other gases contained in air. A survey of literature indicated that palm shell have been used the most as the substrate for CMS production by many researcher in Malaysia [97-101]. Basically, there are three steps involve to prepare the CMS from oil palm wastes which are carbonisation of the wastes, activation of the chars produced and pore modification of the activated carbons to obtain CMS. Activated carbon is produced from carbonaceous source materials such as nutshells, oil palm wastes, peat, wood, coir and lignite. Activated carbon also called activated charcoal is a form of carbon that has been processed to make it extremely porous and have a very large surface area, thus available for adsorption or chemical reactions. Activated carbon can be produced by either physical reactivation or chemical activation. In physical reactivation, the precursor is developed into activated carbons using gases by carbonization and/or oxidation process. For chemical activation, prior to carbonization, the raw material is impregnated with certain chemicals such acid, base or salt [102]. According to a study, the optimum conditions for preparing activated carbon from EFB for adsorption of 2,4,6-TCP were found as follows : activation temperature of 814°C, $CO_2$ activation time of 1.9h and IR of 2.8, which resulted in 168.89 mg/g of 2,4,6-TCP uptake and 17.96% of activated carbon yield [103].

Biochar is commonly defined as charred organic matter, produced to abate the enhanced greenhouse effect by sequestering carbon in soils and improve soil properties. Biochar is a stable carbon compound that can be kept in the ground for a long time, until thousands of years. Biochar is created when biomass is heated to temperatures between 300 and 1000°C, under low or zero oxygen concentrations. Universiti Putra Malaysia (UPM) with the collaboration of Nasmech Technology Sdn Bhd have successfully built a plant producing biochar from EFP and also the first large-scale biochar production plant in the region. They have constructed a carbonator - driven plant to produce the biochar from residue materials including the EFB about 20 tonnes daily [104].

Besides converting dried oil palm wastes into various value added products, it also have potential as a source of renewable energy. Utilization of oil palm wastes as a source of energy will bring other environmental benefit like reduction in $CO_2$.emissions. The greenhouse gases that are present in the atmosphere include water vapor, $CO_2$, methane and ozone, and the increase of greenhouse gases primarily $CO_2$ is the major cause for global warming. Oil palm wastes such as fiber, shell and EFB can be used to produce steam for processing activities and for generating electricity [105]. At present, there are more than 300 palm oil mills operating with self-generated electricity from oil palm wastes. The electricity generated is for their internal consumption and also sufficient for surrounding remote areas [106].

A cement company in Malaysia had used palm shell as fuel in the boiler and they found they the emissions of $CO_2$ can be reduced by 366.26 thousand metric tonnes in the year 2006 alone [107]. Hence, the emission of $CO_2$ in Malaysia can be decreased significantly if all industries in Malaysia can replace or partially replace fossil fuel with oil palm wastes to generate energy without degrading the environment.

Hydrogen is a synthetic fuel, which can be obtained from fossil fuels, nuclear energy and renewable energy sources such as oil palm wastes. In almost any application replacing fossil fuels, hydrogen may be used as fuel especially as feedstock for synthesis of clean transportation fuels or as a gaseous fuel for power generation [108,109]. Gasification is one of the technologies for producing hydrogen. Oil palm wastes such as EFB, fiber, shell, trunks and fronds can be used for gasification [109,110]. The benefits of using hydrogen as transportation fuel are higher engine efficiencies and zero emissions [111]. However, production of hydrogen from oil palm wastes is still at the early stage of research in Malaysia.

# CONCLUSIONS

Malaysia is one of the world's primary palm oil producers and has been taking steps to promote the use of renewable energy. The utilization of renewable energy resources, in particular oil palm wastes is strategically viable as it can contribute to the country's sustainability of energy supply while minimizing the negative impacts of energy generation on the environment. It will help the government to achieve its obligation to prolong the fossil fuel reserves. The efficient use of oil palm biomass other than the palm oil itself for food consumption is a promising route to obtain more energy from oil palm plantations. It will also solve the agriculture disposal problem in an environmental friendly manner while recovering energy and higher value chemicals for commercial applications like bio-fuel, coal replacement, building products and many others. The current principle adopted in Malaysia is a cost pass-through mechanism for electricity generation which is the same principle adopted for renewable power generation. This method would result in a small increase in the price of electricity paid by electricity consumers, but at the same time, the consumers may benefit from revenues derived from renewable energy generation. Although this effort pales in comparison to other countries which had become leaders in renewable energy growth, the acceptance of this form of renewable energy contribution calls for a paradigm shift among the people in the realm of sustainable energy. In general, the maturity of the country is marked by an acceptance of the need for the country to wean reliance on a depleting and environmentally damaging fuel source.

# ACKNOWLEDGEMENTS

The authors would like to acknowledge the three research grants provided by Universiti Sains Malaysia, Penang (1001/PFIZIK/814087, 304/PFIZIK/6310087, 304/PFIZIK/6310073) that has made this research possible.

# REFERENCES

1.    Yusof BasironPalm Oil and Its Global Supply and Demand Prospects. Oil Palm Industry Economic Journal 2002

2.    Malaysian Palm Oil Industry Performance 2008. Global Oils & Fats Business Magazine 6 1 Jan-March), 2009 Available on line at <http://theoilpalm.org/wp-content/themes/oilpalm/pdf/GOFB%206-1%20Malaysian%20Palm%20Oil%20Industry%20Performance%202006.pdf>, (accessed 25 June 2012).

3.    World Growth Palm Oil Green Development Campaign: Palm Oil- The Sustainable Oil a Report by World Growth September 2009. Available on line at <http://www.worldgrowth.org/assets/files/Palm_Oil.pdf>, (accessed 25 June 2012

4.    Teoh Cheng HaiLand Use and the Oil Palm Industry in Malaysia, A bridged report produced for the WWF Forest Information System Database under project MY 0057Policy Assessment of Malaysia Conservation Issues, Kinabatangan, November 2000

5.  K. O. Lim, The Future Energy Potential Of Replanting Oil Palm Tree In Malaysia. Renewable Energy 198687685

6.  United Nations Development ProgrammeMalaysia Generating Renewable Energy from Oil Palm Wastes, UNDP, Kuala Lumpur. 2007

7.  Government of MalaysiaThe Oil Palm in Malaysia. Report published by the Ministry of Agriculture and Cooperatives, Kuala Lumpur, Government of Malaysia. 1966

8.  Corley RHVHardon JJ, Wood BJ. Developments in Crop Science 1, Oil Palm ResearchAmsterdasm:New York:Elsevier Scientific Publishing Company. 1976

9.  Hartley CWSOil Palm (Elaeis Guineensis Jacq.) 2nd ed., London:Longman. 1977

10. K. O. Lim, Zainal Alauddin ZA, Gulam Abdul Qadir, Abdullah MZ. Energy Productivity of Some Plantation Crops in Malaysian and the Status of Bio Energy Utilization. To appear in Proceeding of WREC. 1999

11. H. Kamaruddin, Abdul Halim H, Riazuan R, Mohd. Zain J. Improvement of Oil Palm Fronds CTM-Pulp Brightness-A Preliminary Study, in Proceeding of Seminar on Oil Palm Trunks and Other Palmwood Utilization (Oil Palm Tree Utilization Committee of Malaysia), Kuala Lumpur, Malaysia. 1991

12. Mohamad HusinAbd. Halim Hj. Hassan, Ahmad Tarmizi Mohammed. Availability and Potential Utilisation of Oil Palm Trunks and Fronds up to the year 2000PORIM occasional paper. 20171986

13. M. A. Noor, F. N. Ani, and K. S. Kannan, Renewable Energy Scenario In Malaysia, 2nd International Seminar on Renewable Energy for Poverty Alleviation at IEB, Dhaka, Bangladesh, 1999

14. Foreign Agricultural ServiceUSDA, Oil Seeds Circular, Growing Industrial Use of Vegetable Oil Expected to Impact EU Oilseeds and Products Trade. October 2005

15. USDAOil Crops Outlook, Indonesian Production and Exports of Palm Oil Are Expanding Rapidly. September 2005

16. da Costa RCPotential for Producing Bio-Fuel in the Amazon Deforested AreasBiomass and Bioenergy2004405 EOF415 EOF

17.  E. A. Baryeh, Effects of Palm Oil Processing Parameters on YieldJournal of Food Engineering1 EOF6 EOF

18.  Department of Statistics MalaysiaMalaysian External Trade Statistics, September 2005

19.  Foreign Agricultural ServiceUSDA, Oil Seeds Circular, Table 9. October 2005

20.  Review of Annual Production Of Oil Palm Products : 19752005 2005. Malaysian Oil Palm Board (MPOB). <http://econ.mpob. gov.my/economy/Production1_05.htm>, (accessed 5 July 2009).

21.  Energy Information Administration (EIA)US Department of Energy, Country Analysis Brief Malaysia, Update January 2005

22.  Review of the Malaysian Oil Palm Industry 2006Malaysian Oil Palm Board (MPOB).

23.  A Summary On The Performance if The Malaysian Oil Palm Industry. (2008). Report by Malaysian Oil Palm Board. <http:// econ.mpob.gov.my/economy/Performance-130109.html>, (accessed 1 August 2009).

24.  Overview of the Malaysian Oil Palm Industry. (2002). Report by Malaysian Oil Palm Board. <http://econ.mpob.gov.my/review 2002.html>, (accessed 6 May 2009

25.  Radzi Abas MOros DR, Simoneit BRT. Biomass Burning as The Main Source of Organic Aerosol Particulate Matter in Malaysia During Haze Episodes.Chemosphere200455108995

26.  Y. Basiron, and C. K. Weng, The Oil Palm and Its SustainabilityJournal of Oil Palm Research20041 EOF10 EOF

27.  MB Wahid, , Abdullah SNA, Henson IE. Oil Palm- Achievements and Potential. In: New Directions for a Diverse Planet, Proceedings of the 4th International Crop Science Congress, 26 Sep- 1 Oct 2004, Brisbane, Australia. <www.cropscience.org. au>, (accessed 21 September 2009).

28.  Mahlia TMIAbdulmuin MZ, Alamsyah TMI, Mukhlishien D. An Alternative Energy Source from Palm Waste Industry For Malaysia and Indonesia,Energy Conversion and Management200142210918

29.  S. Prasertsan, and P. Prasertsan, Biomass Residues from Palm Oil Mills in Thailand: An Overview on Quantity and Potential UsageBiomass and Bioenergy1996387 EOF395 EOF

30. S. Yusoff, Renewable Energy from Palm Oil- Innovation on Effective Utilization of Waste. Journal of Cleaner Production 2006148793

31. Z. Husain, Z. A. Zainal, M. Z. Abdullah, Analysis of Biomass-Residue-Based Cogeneration System in Palm Oil MillsBiomass and Bioenergy20032411724

32. A. L. Ahmad, S. Ismail, S. Bhatia, Water Recycling from Palm Oil Mill Effluent (POME) using Membrane TechnologyDesalination20031578795

33. Hoi Why Kong, Koh Mok Poh. Newsletter of Wood and Non-Wood Products Utilization Branch (FOPW) of the FAO Forest Products Division, Issue 9 December 2001. <http://www.fao.org/DOCREP/003/y3198E/y3198e00.htm> (accessed 19 September 2002).

34. H. Y. Kong, Current status of biomass utilization in Malaysia. Forest Research Institute Malaysia; 2000115

35. C. S. Goh, K. T. Tan, K. T. Lee, S. Bhatia, Bio-Ethanol From Lignocellulose: Status, Perspectives and Challenges in Malaysia. Bioresource Technology2010July (13)):4834 EOF41 EOF

36. Z. Hussain, Z. Zainac, Z. Abdullah, Briquetting of Palm Fibre and Shell from The Processing of Palm Nuts to Palm Oil. Biomass and Bioenergy 2002

37. H. H. Yeoh, The Potential Of Producing Liquid Fuel From Oil Palm Wastes. M.Sc. thesis. Universiti Sains Malaysia, Penang, Malaysia; 1999

38. B. Lim, The New Straits Times 2000Thursday, December 28, 2000.

39. A. B. Hamdan, A. M. Tarmizi, D. Mohad, Tayeb. Empty Fruit Bunch Mulching and Nitrogen Fertilizer Amendment: The Result Effect On Oil Palm Performance And Soil Properties, PORIM Bulletin 1998Palm Oil Resesrch Institute of Malaysia.

40. A. Z. Zain, Koh Mok Poh, Mohd. Yusoff MN, Khoo Kean Choon, Mohd. Nasir N. Commercial Utilization of Oil Palm Empty Fruit Bunches, in Processing of the Third National Seminar on Utilisation of Oil Palm Tree and Other Palms, Forest Research Institute Malaysia, Kuala Lumpur; 1994

41. Abu Hassan Olshida M. Status of Utilization of Selected Fibrous Crop Residues and Animal Performance with Emphasis on Processing of Oil Palm Fronds (OPF) For Ruminant Feed in Malaysia. Tropical Agriculture Research Center TARS 25Ministry of Agriculture, Forestry and Fisheries, Tsukuba, Japan; 1992

42. A. Takigawa, S. Oshio, Mohd. Jaafar Daud, Abu Hassan Osman. Processing and Utilization of Palm By-product for Ruminant, MARDI-TARC Collaborative Study; 1990

43. Khozirah ShaariKhoo Kean Choon () Introduction. In Khozirah Shaari, Khoo K.C, Abdul Razak Mohd. Ali (eds.) Oil Palm Stem Utilisation 1991: Review of Research, FRIM Research Pamphlet 107Kepong, Forest Research Institute Malaysia; 1991

44. M. Husin, R. Ramli, A. Mokhtar, Hassan WHW, Hassan K, Mamat R, Aziz AA. Research and Development of Oil Palm Biomass Utilization in Wood-based Industries. Palm Oil Developments 2002

45. N. R. Menon, Z. A. Rahman, and N. A. Bakar, Empty Fruit Bunches Evaluation: Mulch in Plantation vs. Fuel for Electricity Generation. Oil Palm Industry Economic Journal 2003321520

46. M. Z. Jaafar, W. H. Kheng, N. Kamaruddin, Greener Energy Solutions for A Sustainable Future: Issues and Challenges for MalaysiaEnergy Policy20033111106172

47. I-R. a. z. i A. Fakhru, Noor MJMM. Treatment of Palm Oil Mill Effluent (POME) with the Membrane Anaerobic System (MAS) Water Science and Technology 1999159 EOF

48. K. K. Chin, S. W. Lee, H. H. Mohammad, A Study of Palm Oil Mill Effluent Treatment Using a Pond SystemWater Science and Treatment 1996119 EOF

49. F. N. Ani, M. N. Islam, Liquid Oil from Fluidised Bed Pyrolysis of Rice Husk Waste and Its Characterisation, RERIC International Energy Journal 1998

50. A. C. Lua, J. Guo, Preparation And Characterisation of Chars from Oil Palm Waste, Carbon 1998

51. K. O. Lim, Production of Solid Biofuels from Biowastes, in Proceedings of The Second International Energy Conference, Kuala Lumpur, Energy from Biomass Residues, Forest Research Institute Malaysia; 1992

52. T. Y. Chew, B. G. Yeoh, Palm Oil Effluent As A Source of Bioenergy, in Proceedings National Symposium on Oil Palm By-Products for Agro-Based Industries, Kuala Lumpur; 1985

53. P. F. Akmar, P. Elham, Ethanol Fermentation from Oil Palm Trunk Hydrolyzate, in Oil Palm Trunk and Other Palmwood Utilization, Proceeding of a National Seminar, Kuala Lumpur, Malaysia; 1991

54. A. W. Mohammad, W. T. Yeong, Md Jahim J, Anuar,N. Palm Oil Mill Effluent (POME) Treatment and Bioresource Recovery Using Ultrafiltration Membrane: Effect of Pressure One Membrane Fouling, Biochemical Engineering Journal 200835330917

55. A. N. Ma, A Novel Treatment for Palm Oil Mill Effluent, Palm Oil Research Institute of Malaysia (PORIM) 19952920112

56. T. K. Hwang, S. M. Ong, C. C. Seow, H. K. Tan, Chemical Composition of Palm Oil Mill Effluent, Planter 19785474956

57. Habib MAB Yusoff FM, Phang SM, Ang KJ, Mohamed S. Nutritional Values of Chironomid Larvae Grown in Palm Oil Mill Effluent and Alga Culture.Aquaculture 199715895105

58. P. Agamuthu, E. L. Tan, Digestion of Dried Palm Oil Mill Effluent by Cellulomonas Species. Microbios Letters 19853010913

59. S. Gurmit, Lim Kim H, Teo L, David Lee, K. Oil Palm and the Environment a Malaysian Perspective,Malaysian Oil Palm Growers Council 19991253

60. 'Palm biomass power plants in Malaysia'. Biopact article. Sunday, August 06, 2006. <http://biopact.com>, (accessed 23 September 2009).

61. The world's most productive ethanol crop is oil palm'. Biopact article. June 21st, 2006. <http://biopact.com>, (accessed 23 September 2009

62. O. U. Esther, Anaerobic Digestion of Palm Oil Mill Effluent and Its Utilization as Fertilizer for Environmental ProtectionRenewable Energy1997102132914

63. I-R. a. z. i. Fakhru, Yassin AAA, Lyuke SE, Ngan MA, Morimoto M. Bio-hydrogen Synthesis from Wastewater by Anaerobic Fermentation Using MicrofloraInternational Journal of Green Energy2005238796

64. Bumibiopower Methane Extraction and Power Generation'. Project report by Mitsubishi Securities Clean Energy Finance

Committee. (January 2004). <http://dnv.com/certification/.../ PDD_BumibiopowerBiomass_2004-03-19.pdf> (accessed 5 August 2009

65. Wikipedia. The free encyclopedia: pyrolysis oil. <http:// en.wikipedia.org/wiki/Bio-oil> (accessed 23 April 2011).

66. A. V. Bridgwater, Peacocke GVC. Fast Pyrolysis Processes for BiomassSustainable and Renewable Energy Reviews 200041173

67. D. Chiaramonti, A. Oasmaa, Y. Solantausta, Power Generation using Fast Pyrolysis Liquids from BiomassRenewable and Sustainable Energy Reviews2007116105686

68. A. V. Bridgwater, D. Meier, D. Radlein, An Overview of Fast Pyrolysis of BiomassOrganic Geochemistry19993012147993

69. Bumibiopower Biomass Power Plant Project Malaysia. Project Design Document for Small-Scale SDM Activity, Version 02, March 2006. <http://www.dnv.com/certification/climatechange/ Upload/PDD_Bumibiopower%20Biomass_2004-03-19.pdf > (accessed 23 December 2007).

70. Nurul Islam MZailani R, Ani FN. Pyrolytic Oil from Fluidised Bed Pyrolysis of Oil Palm Shell and Its Characterisation, Renewable Energy 1999

71. Kawser MDJFarid NA. Thermochemical Processing of Rubber Waste to Liquid FuelJournal of Plastics, Rubber and Composites Processing and Applications 20002921004

72. F. N. Ani, M. N. Islam, Liquid Oil from Fluidised Bed Pyrolysis of Rice Husk Waste and its Characterisation, RERIC International Energy Journal 19982015565

73. F. N. Ani, M. N. Islam, Pyrolytic Recycling of Agro-Industrial Solid Wastes in MalaysiaJournal of the Institute of Energy199871558

74. F. Abnisa, Wan Daud WMA, Husin WNW, Sahu JN. Utilisation Possibilities of Palm Shell as a Source Of Biomass Energy In Malaysia By Producing Bio-Oil In Pyrolysis Process. Biomass and Bioenergy 201135186372

75. N. Abdullah, An Assessment of Pyrolysis for Processing Empty Fruit BunchesPhD thesis, Aston University, Birmingham, UK; 2005

76. M. A. Sukiran, C. M. Chin, Abu Bakar NK. Bio-oils from Pyrolysis of Oil Palm Empty Fruit BunchesAmerican Journal of Applied

Sciences20096586975

77.  Kawser MDJFarid Nasir A. Oil Palm Shell as a Source of PhenolJournal of Oil Palm Research20001218694

78.  'Genting Group Unveils Malaysia's First Commercially Produced Bio-oil Using Breakthrough Technology'Report by Genting Group, Press Release, 21st August, 2005

79.  Venderbosch RH, Gansekoele E, Florijn JF, Assink D. Pyrolysis of Oil Palm Residue in Malaysia. BTG Biomass Technology Group BV, Ng HY, Genting Bio-oil BHD. <http://www.pyne.co.uk/docs/PyNe%20Section1.pdf;2005>, (accessed 3 April 2011

80.  BTG-BTL website. Biomass To Liquid:Pyrolysis Oil-The Sustainable Alternative. <http://www.btg-btl.com/index2.php> (accessed 2 April 2011).

81.  Poster A. Malaysia Palm Oil Council. Palm Oil Industry- a Learning Experient. 2006. <http://www.mpoc.org.,y/>, (accessed January 2009

82.  'Long Term Benefit from Oil Palm Biomass'Bernama, The Malaysian National News Agency. Press Release 16th August, 2001

83.  M. Husin, R. Ramli, A. Mokhtar, Wan Hassan WH, Hassan K, Ropandi, M, Abdul Aziz A. Research and Development of Oil Palm Biomass in Wood Based Industries. Malaysian Palm Oil Board Bulletin. Palm Oil Development 36. 2005

84.  912240421Sumanthi, S., Chai, S.P. and Mohamed, A.R. Utilization of Oil Palm as a Source of Renewable Energy in Malaysia. Renewable and Sustainable Energy Reviews 2008; 12(9): 2404-21

85.  Malaysian Timber Council (MTC). Kuala Lumpur, Malaysia (1992-2004). <http://www.mtc.com.my/news/pr33.htm>, (accessed 7 January 2009).

86.  Wood Based Industry. Report by Malaysia Industrial Development Authority (MIDA). http://www.mida.gov.my/en_2 index.php?page=wood-based-industry; (accessed 7 April 2011).

87.  MIDA e-News: Agro-Bio Fibre expands production, ties up with FRIM on HDF. Report by Malaysia Industrial Development Authority (MIDA). <http://www.mida.gov.my/en_2 index.php?m act=News,cntnt01,detail,0&cntnt01articleid=234&cntnt01retur

nid=388>, (accessed 7 April 2011).

88. Ministry of Plantation Industries and Commodities. Speech by Y.B. Dato' Seri Lim Keng Yaik in the Opening Ceremony of the Seminar on Oil Palm Biomass: Opportunities for Commercialization (2001). <www.kppk.gov.my/index.php?option=com_content&task=view&id=86&Itemid=4653 k>, (accessed 9 September 2009).

89. S. H. Shuit, K. T. Tan, K. T. Lee, A. H. Kamaruddin, Oil Palm Biomass as a Sustainable Energy Source: Malaysian Case Study, Energy 2009349122535

90. OA Hassan, , M Ishida, , Shukri Mohd I, Tajuddin ZA. Oil Palm Fronds as a Roughage Feed Source For Ruminants in Malaysia FFTC for the Asia and Pacific Region 1994. <http://www.agnet.org/library/eb/420/;1994>, (accessed 6 September 2009).

91. 'Oil Palm Frond Based Ruminant Pellet'. (2008). Report by MARDI, <http://www.mardi.my/main.php?Content=home&FolderID=157&CurLocation=131>, (accessed 7 September 2009).

92. N. F. Zainudin, K. T. Lee, A. H. Kamaruddin, S. Bhatia, A. R. Mohamed, Study of Adsorbent Prepared From Oil Palm Ash (OPA) for Flue Gas Desulphurization. Separation and Purification Technology 2005455060

93. A. R. Mohamed, N. F. Zainudin, K. T. Lee, A. H. Kamaruddin, Reactivity of Adsorbent Prepared From Oil Palm Ash (OPA) for Flue Gas Desulphurization : Effect of SO2 Concentration and Reaction Temperature. Studies in Surface Science and Catalysis 200615944952

94. S. M. Nomanbhay, K. Palanisamy, Removal of Heavy Metal from Industrial Wastewater Using Chitosan Coated Oil Palm Shell CharcoalElectronic Journal of Biotechnology 2005April 15):44-53.

95. A. B. Nasrin, A. N. Ma, Y. M. Choo, S. Mohamad, M. H. Rohaya, A. Azali, et alOil Palm Biomass as Potential Substitution Raw Materials for Commercial Biomass Briquettes ProductionAmerican Journal of Applied Science 20085317983

96. Hoi Why KongWood and Charcoal Briquetting in Malaysia. In: Proceedings of International Workshop on Biomass Briquetting, (eds P.D.Grover and S.K Mishra). 19956874FAO Regional Wood Energy Development Programme. 1995.

97. M. Ahmad, Wan Daud W, Aroua M. Adsorption Kinetics of Various Gases In Carbon Molecular Sieves (CMS) Produced From Palm ShellColloids and Surfaces A 20083121315

98. J. Tan, F. Ani, Carbon Molecular Sieves Produced From Oil Palm Shell for Air SeparationSeparation and Purification Technology2004354754

99. M. Ahmad, Wan Daud W. Aroua M. Synthesis of Carbon Molecular Sieves from Palm Shell by Carbon Vapor DepositionJournal of Porous Matter 2007143939

100. D. Adinata, Wan Daud W, Aroua M. Production of Carbon Molecular Sieves From Palm Shell Based Activated Carbon by Pore Sizes Modification With Benzene For Methane Selective SeparationFuel Process Technology 200788599605

101. W. Daud, M. Ahmad, M. Aroua, Carbon Molecular Sieves from Palm Shell: Effect of the Benzene Deposition Times On Gas Separation PropertiesSeparation and Purification Technology20075728993

102. Wikipedia. The free encyclopedia: Activated Carbon. <http://en.wikipedia.org/wiki/Activated_carbon#cite_note-0>, (accessed 3 April 2011).

103. B. H. Hameed, Tan IAW, Ahmad AL. Preparation of Oil Palm Empty Fruit Bunch-Based Activated Carbon For Removal Of 246Trichlorophenol: Optimization using Response Surface Methodology.Journal of Hazardous Materials2009

104. Bernama. Biochar Malaysia: UPM-NASMECH Effect of Producing EFB Biochar : World's First. 7 January 2010. <http://biocharmalaysia.blogspot.com/2010/01/upm-nasmech-effort-of-producing-efb.html>, (accessed 4 April 2011).

105. Keck Seng (M) Berhad. Biogas utilization. 2001. <http://www.biogen.org.my/bris/BioGen/Tech/(b)Project/technology(b)2.pdf>, (accessed 4 April 2011).

106. NA Ludin, , MA MBakri, , M Hashim, , B Sawilla, , NR Menon, , H Mokhtar, . Palm Oil Biomass For Electricity Generation In Malaysia. Pusat Tenaga Malaysia, Malaysia Palm Oil Board, SIRIM Berhad. 2004. <http://www.biogen.org.my/bris/BioGen/Tech/(d)Documents/technology(d)7.pdf>, (accessed 5 April 2011).

107. M. Dit, Palm Kernel Shell (PKS) Is More than Biomass for Alternative Fuel after 2005. Proceedings of Chemistry and Technology Conference. 2007cdm.unfccc.int/Projects/DB/DNV-CUK1182238337.59/ReviewInitialComments/6MHGPZIM56I19DT6NGL30G9OGNTIOK>, (accessed 3 April 2011).

108. F. Barbir, Transition to Renewable Energy Systems with Hydrogen as an Barbir F. Transition to Renewable Energy Systems with Hydrogen as an Energy Carrier. Energy 2009; .CarrierEnergy 20093430812

109. A. Azali, A. B. Nasrin, Y. M. Choo, N. M. Adam, S. M. Sapuan, Development of Gasification System Fuelled With Oil Palm Fibres and Shells. American Journal of Applied Science 20052005725

110. T. L. Kelly-yong, K. T. Lee, A. R. Mohamed, S. Bhatia, Potential of Hydrogen From Oil Palm Biomass as a Source Of Renewable Energy WorldwideEnergy Policy2007355692701

111. H. J. Neef, International Overview of Hydrogen and Fuel Cell ResearchEnergy20093432733

# Chapter 7

# Lignocelluloses Feedstock Biorefinery as Petrorefinery Substitutes

Cheng Hongbin[1, 3] and Wang Lei[2, 3]

[1]Department of Process Engineering, Stellenbosch University, Stellenbosch, South Africa, South africa

[2]Department of Life Science, Imperial College London, London, UK

[3]New China Times Technology Ltd, China

# INTRODUCTION

## Lignocelluloses Feedstock (LCF) Biorefinery

### *Background*

The material needs from our society are reaching the crisis point, as the demand for resources will soon exceed the capacity of the present fossil resource based infrastructure [1]. Currently, fossil-based energy resources, such as petroleum, coal, and natural gas, are responsible for about three quarters of the primary energy consumption in our world. While decreasing crude-oil reserves, enhanced demand for fuels worldwide, increased climate concerns about the use of fossil-based energy carriers, and political commitment, the focus has recently turned to develop the utilization of renewable energy resources [2]. Gullón *et al.* [3] described the variety of problems on present social, economic and technological situation, which including: the fear for a shortening of the supplies of basic resources, as the population growth; the increasing *per capita* demands of the developing economies for goods and energy, derived from the increasing purchase power of the population; environmental challenges, especially those related to effects of greenhouse gas emissions (emphasis on $CO_2$) on the global climate; the national security issues surrounding reliance on imported oil [4].

On our market, nowadays, there are more than 2500 different oil-based products. The petroleum crisis of the 1970s resulted in a shift from total reliance on fossil resources and simultaneously triggered research into biomass based technologies. As a result of the oil crisis, renewable resources became a popular phrase [5]. Currently, the most of energy requirements in the world are still met by fossil fuels. The limited deposits of these fossil fuels coupled with environmental problems have prompted people to look for sustainable resources as alternatives to meet the increasing energy demand. Bio-energy production has the advantage of forming smaller amounts of greenhouse gases compared to the conversion of fossil fuels, as the carbon dioxide generated during the energy conversion is consumed during subsequent biomass re-growth [6]. However, simply providing sustainable and non-polluting energy will not be enough. In our life, clothes, shelter, tools, medications and so on are all, to a greater or lesser degree, dependent on organic carbon. As fossil-based resources will be replaced, new sources of organic carbon have be found or alternate applications and processing of existing sources must be developed. The challenge is to find replacements not only for current usage, but also for the even future greater energy consumption, with a likely concomitant increase in biomass demand for manufacturing [7].

## *What is Biorefinery?*

The core aim for biorefineries is to produce both high-volume liquid fuels and high-value chemicals. As petroleum refinery uses petroleum as the major input and processes it into many different products, the term 'biorefinery' has been coined to describe the processing complexes that will use biomass as feedstocks to produce a wide spectrum of chemicals, fuels and bio-based materials, that can be used as industrial intermediates or sold directly to consumers [1, 8, 9]. Biorefineries have been considered as the key for access to an integrated production of chemicals, materials, goods, fuels and energy of the future [10]. As oil prices continue to rise and biorefining technology matures, biorefineries are playing an increasingly major role in the global economic system, with the potential to ultimately replace petroleum refineries as the world's principal method of fuel generation.

## *Lignocelluloses Feedstock (LCF) Biorefinery*

The largest organic carbon reservoir in our world is the biomass - plants and algae. Each year, plants fix approximately 90 billion tons of $CO_2$, most of this as wood [11]. Lignocelluloses are the natural combination of cellulose, hemicelluloses and lignin. It's the raw material for potential conversion to energy fuels and chemical feedstock for manufacturing. LCF biorefinery has been drfined as one of the so-called phase-III biorefinery concepts which are characterized by the ability to use a variety of resources by different routes to generate multiple products [12].

A LCF biorefinery uses lignocellulosic biomass, including forestry residue, agricultural residue, yard waste, wood products, animal wastes, etc. Initially, plant material is cleaned and broken down into the three main fractions (hemicellulose, cellulose, and lignin) by chemical digestion or enzymatic hydrolysis. Hemicellulose and cellulose can be produced by alkaline and acid. Lignin can also be further broken down with enzymes. The hemicellulose and cellulose are sugar polymers, which can be converted to their component sugars through hydrolysis. A hemicellulose is a polymer that contains five-carbon sugars (usually D-xylose and L-arabinose), six-carbon sugars (D-galactose, D-glucose, and D-mannose), and uronic acid. Cellulose is a polymer of only glucose. The hydrolysis process of hemicelluloses and cellulose result in the aforementioned sugars [13].

The LCF Biorefinery is a promising alternative due to the abundance and variety of available raw materials and the good position of the conversion products on the market [14]. Its profitability is also dependent on the technology employed to alter the structure of lignocellulosic biomass in order to produce high value co-products from its three main fractions *i.e.* cellulose, hemicellulose, and lignin [15].

Currently the main feedstock for biorefineries is still based on starch. The practiced technologies in fuel ethanol industry are primarily based on the fermentation of sugars derived from starch and sugar crops, which are quite mature with little possibility of process improvements. However, using starch and sugar crops to produce ethanol also has been questioned since it draws its feedstock from a food stream. Lignocellulosic biomass is a more promising renewable resource as it is available in large quantities and does not compete with food or feed. Lignocellulosic biomass is a renewable resource that stores energy from sunlight in its chemical bonds, with great potentials for the production of affordable fuel ethanol [16, 17]. Its main obstacle for a major breakthrough is the high production costs for bioenergy products.

On the other hand, lignocellulosic biomass-derived products can significantly reduce green house gas emissions, compared to fossil-based products. Also, many common petrochemicals could be obtained with lower greenhouse gas emissions from bio-based feedstocks. The maturity and economics of the conversion processes and logistics is a major challenge for lignocellulosic biomass [18].

## The Main Goal of Biorefinery

With, implementing innovative, environmentally sound and cost-effective production technologies for a variety of products, the integrated biorefinery is increasing the availability and use of bioenergy and bio-based products. The main objective of a biorefinery is to produce high-value low volume and low-value high-volume products by a series of producing processes. The processes are designed to maximize the valued products while minimizing the waste streams by converting low-value high-volume intermediates into energy. The high-value products can enhance the profitability, and the high-volume fuels will help to meet the global energy demand. The power produced from a biorefinery can also help to reduce the overall cost. Figure 1 shows the elements of a biorefinery, in which biomass is used to produce various useful products such as fuel, power, and chemicals by biological and chemical conversion processes [13].

Traditionally, the matured biorefinery pathways include bioconversion (aerobic and anaerobic digestion) and chemical conversion (bio-pulping). There are two most promising emerging biorefinery platforms. One is the sugar platform and the other is the thermo-chemical platform (syngas platform). In sugar biorefineries platform, biomass will be broken down into different types of component sugars for fermentation or other biological processing into various fuels and chemicals. In thermo-chemical biorefineries platform, biomass will be synthesized hydrogen and carbon monoxide or pyrolysis oil, the various components of which could be directly used as fuel [19].

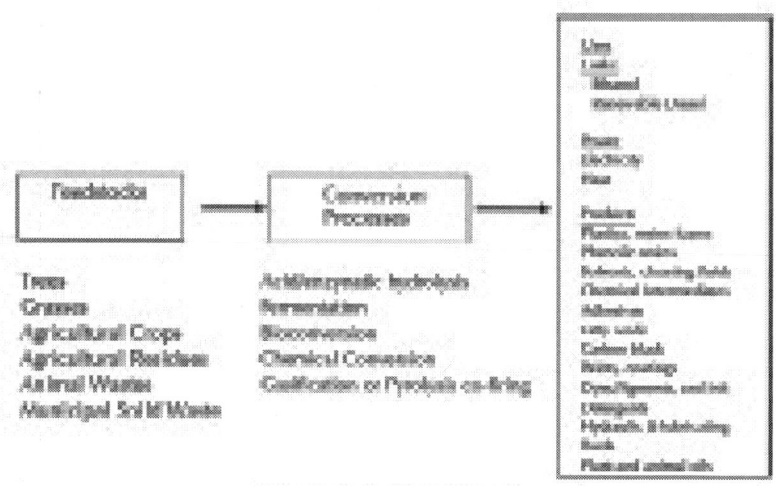

**Figure 1:** Simple procedure for three-step biomass-process-products [13].

## *Disadvantages*

It is very important to increase the reaction rates, as slow reactions rates is one of the main disadvantages for biological conversions in biorefinery processes. Another disadvantage is the often low product concentrations, which means the high product recovery costs with existing technology. The lower yields of targeted products is often found in some multiple products systems [20]. Therefore the biorefinery processes to become an actual alternative to fossil fuels and petroleum-derived products, biorefinery processes must be competitive and cost-effective [21].

# BIOFUELS

As an important category of bioenergy, biofuel is a type of fuel which is biologically derived from biomass. The biofuels, which include liquid, solid biofuels and various biogases, can replace the conventional petroleum or petroleum derived products. Many biological reactions involved in biofuels production are at mild conditions, can offer relatively high products yields and generally result in low levels of contamination to the environment. The modern application of biological transformations, known as biotechnology, is also an evolving field that has great promise for substantial improvements and significant cost reductions. In this section, several liquid and gases biofuels are introduced e.g. (1) bioethanol, biobutanol, and biodiesel which can replace the gasoline used as transportation fuels; and (2) biogas, which is produced from anaerobic digestion of biomass as a substitute for natural gas either for industrial applications or for transportation.

# Bioethanol

Bioethanol is a promising transport fuel alternative to gasoline because it has higher oxygen content and no sulphur or nitrogen when compared with gasoline [22]. Currently, the blends E5 and E10 that consist of 5% (v/v) and 10% (v/v) ethanol respectively, have a widespread usage since these blends can supply the existing vehicular fleet without major changes to engines. High bioethanol blends (E100, E95 and E85) require modified or dedicated vehicles.

Bioethanol can be produced from three types of raw materials: sugars (from sugarcane, sugar beet, molasses, and fruits), starch (from corn, cassava, potatoes, and root crops), and cellulose (from wood, agricultural residues, waste sulphite liquor from pulp and paper mills). Among the three main types of raw materials, cellulose contained in lignocellulosic biomass represents the most abundant global source of biomass, which can be utilised for bioethanol production [23]. There are also two approaches for producing bioethanol from lignocellulosic biomass through (1) Biochemical (2) Thermochemical processes.

## *Biochemical Production of Bioethanol*

Figure 2 illustrates the high level technologies for producing bioethanol from these various biomass feedstocks. Typically, the common steps for biologically producing bioethanol from different feedstocks are fermentation and distillation. For the first generation (1G) bioethanol production, the sugar extracted from sugar-rich crops and that from starch digestion by amylases or acids is directly fermented to bioethanol. To convert lignocellulosic biomass into second generation (2G) bioethanol, an additional step of pre-treatment is usually required.

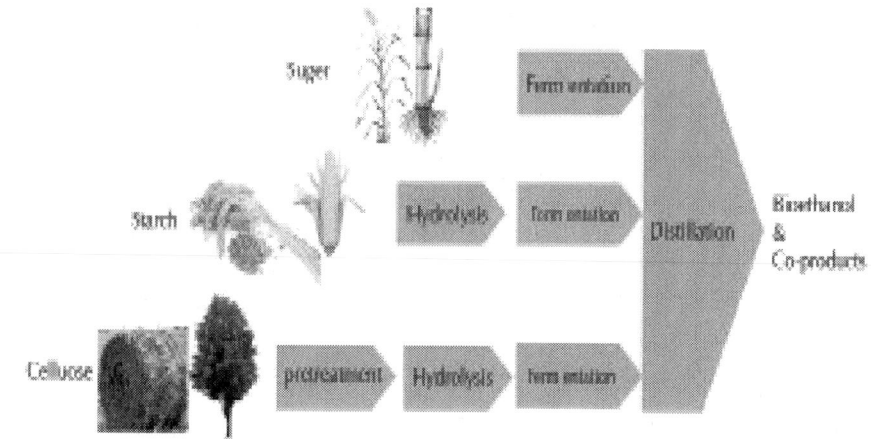

**Figure 2:** Technologies required producing bioethanol from biomass. [24].

A wide variety of lignocellulosic feedstocks are potentially available for bioethanol production such as wood, grass, agricultural waste and MSW (municipal solid waste). Their physical structures and chemical compositions are different; therefore technologies applied for bioethanol production can be diverse. In addition to the main product bioethanol, co-products are also usually produced, such as heat and electricity generated by burning lignin-rich residue from fermentation and also, potentially, a wide range of high value-added chemicals like acetic acid, furfural and hemicellulose sugar syrup and the low molecular weight lignin.

General technologies required for biologically producing 2G bioethanol include (1) pre-treatment, (2) enzymatic hydrolysis, (3) fermentation, and (4) distillation.

Pre-treatment is applied to enhance the accessibility of enzyme to biomass by increasing available biomass particle surface area for enzyme to attack. This can be achieved by partially removing lignin and/or hemicellulose, changing the structure of biomass fibres to decrease cellulose crystallinity and its degree of polymerization. The current available pre-treatment methods can be classified as mechanical, chemical and biological. Table1 summarised some typical pre-treatment methods and their characterisations. Pre-treatment has been viewed as the most expensive step in the biologically production of bioethanol. Therefore, it is important to assess the economic feasibility of the pre-treatment method in addition to its technology performance. More information about each pre-treatment method can be found in Section 5.

Enzymatic hydrolysis is carried out under mild conditions with potentially high sugar yields and relatively low maintenance costs. Nevertheless, major challenges for cost-effective commercialisation remain, such as the high cost of enzymes, the slow rate of enzymatic reaction and potential inhibition by sugar degradation products from pre-treatments [48]. In enzymatic hydrolysis, cellulose is hydrolysed by a suite of enzymes, including cellulase and β-glucosidase crudely purified from lignocellulose-degrading fungi such as *Trichoderma reesi*, *Trichoderma viride* and *Aspergillus niger.*Cellulase refers to a class of enzymes including endocellulase breaking internal bonds of cellulose, exocellulase cleaving from the free ends of chains produced by endocellulase to form cellobiose (a dimer of glucose), and cellobiase (β-glucosidase) then hydrolysing cellobiose to produce glucose monomers. In addition, most of cellulase mixtures contain hemicellulase that facilitates hemicellulose hydrolysis to assist with the overall effectiveness of enzymatic hydrolysis.

After the enzymatic hydrolysis, sugar monomers can then be fermented to ethanol by micro-organisms (*e.g. Saccharomyces cerevisiae* and *Zymomonas mobilis*). Fermentation has been commercialised in brewery and food manufacturing for centuries and itself is not a complex and expensive process. The challenges regarding fermentation for the bioethanol industry are: (1) to convert pentose (C5 sugar) which cannot be fermented by the conventional yeast efficiently, and (2) to prevent inhibition caused by sugar degradation products from pre-treatments. Research has shown the feasibility of construction and application of genetically engineered yeasts capable of converting both pentose and hexose to ethanol [49]. Further potential lies in using bacteria with the metabolic pathways necessary to ferment all sugars available from lignocellulosic biomass. *Z. mobilis* has shown to be capable of metabolising 95% of glucose, 80% of xylose and 40% of other sugars in corn stover hydrolysate [50]. Metabolic engineered *Geobacillus thermoglucosidasius* has demonstrated an ethanol yield of over 90% of theoretical at temperatures in excess of 60°C [51].

**Table 1:** Chemical pre-treatment methods for lignocellulosic biomass

| Pre-treatment method | Process and conditions | Possible changes in biomass | Disadvantages | Reference |
|---|---|---|---|---|
| Steam explosion | No agent temperature:160-260°C,20-50 bar , 2-5 minutes | Dissolve hemicelluloses Low sugar degradation | Partially degrade hemicellulose | [25-27] |
| Ammonia fibre explosion (AFEX) | Ammonia as agent, 65-90°C, 0.5-3 hours | Change biomass physical structure Enhancing hemicelluloses hydrolysis | Limited effects on soft and hardwood | [28, 29] |
| SO2/ H2SO4explosion | SO2 as agent, 160-220°C, < 2 minutes | Dissolve hemicelluloses effectively for hardwood and agricultural residues | Degradation of hemicelluloses, less effective for softwood | [30, 31] |
| CO2explosion | CO2 as agent, 35°C, 56.2 bar, 10-60 minutes | Interrupt crystalline structure of cellulose | Inefficient for softwood and high capital cost | [32, 33] |
| Hot liquid water | Water as agent, 190-230°C, 45 seconds-4 minutes | Effectively dissolve hemicelluloses Very low degradation | Water recycling prohibitively expensive | [34-36] |
| Dilute acid | H2SO4 as agent , over 160°C, 2-10 minutes | Effectively dissolve hemicelluloses | Needs neutralisation, significant formation of fermentation inhibitors | [37-39] |
| Alkaline | NaOH/ Ca(OH)2/ Ammonia as agent, 70-120°C, 20-60 minutes | Removal of lignin Low hemicelluloses degradation | Costs of reagents and wastewater treatment are high | [40-42] |
| Oxidation | Ca(OH)2+O2/ H2O2as agent, 140°C, 3 hours | Removal of lignin Low hemicelluloses degradation | Costs of reagents and wastewater treatment are high | [43, 44] |

| Organic solvent | Ethanol as agent, 140-200°C, 30-150 minutes | Removal of lignin | Cost of solvent recovery is high | [45, 46] |
|---|---|---|---|---|
| Ionic liquid | Ionic liquid as agent, 120°C, 22 hours | Remove of lignin and hemicellulose | Costs of reagents and long treatment time | [47] |

Bioconversion process configurations, including Separate Hydrolysis and Fermentation (SHF), Simultaneously Saccharification and Fermentation (SSF), Simultaneously Saccharification and Co-Fermentation (SSCF), and Consolidated Bioprocessing (CBP). The SHF has many advantages, such as allowing both enzyme and micro-organisms to operate at their optimum conditions. Also, any accidental failure of enzymatic hydrolysis and fermentation would not affect the other steps. Alternatively the enzymatic hydrolysis may also be combined with fermentation and can thus be carried out simultaneously in a same reactor - this being known as the simultaneous saccharification and fermentation (SSF). During enzymatic hydrolysis, the cellulases are strongly inhibited by hydrolysis products: glucose and short cellulose chains ('end-point' inhibition). SSF can overcome this inhibition by fermenting the glucose to ethanol as soon as it appears in solution. However, ethanol itself inhibits the action of fermenting micro-organisms and cellulase although ethanol accumulation is less inhibitory than high concentrations of hydrolysis products [52]. Nevertheless, SSF operating at the compromised temperature (37-40 °C) has some drawbacks caused by the different optimal temperatures for the action of cellulases (45-50° C) and the growth of microorganisms (typically 28-35°C). One method to overcome this disadvantage is the utilisation of thermo-tolerant fermenting organisms. SSCF is a promising SSF process where the micro-organism co-ferment pentose and hexose to bioethanol. CBP currently becomes the focus of most research efforts to date; it integrates cellulase production, cellulose hydrolysis and fermentation in one step by using an engineered strain [53]. Many studies have been reported in CBP technologies developments recently [54-56].

Nevertheless, other significant efforts are also required to enable future integrated biorefinery. They include (1) promising process designs to integrate energy consumption and minimise the water footprint (2) producing a range of high value added by products, e.g. power, chemicals, and lignin-derived products etc.

## *Thermo-chemical Production of Bioethanol*

The thermo-chemical bioethanol production refers to a series of processes including biomass indirect gasification, alcohol synthesis and alcohol separation as shown in Figure 3.

The biomass is processed and dried by flue gas before being fed to biomass gasifier. The biomass is chemically converted to a mixture of syngas components (*i.e.* CO, $CH_4$, CO and $H_2$ *etc*), tars, and a solid char which is the fixed carbon residual from the biomass. The heat required for endothermic gasification reactions is supplied by circulating hot synthetic olivine 'sand' between the gasifier and combustor. The solid char and 'sand' from the gasifier are separated by cyclones and then sent to a char combustor where the char is oxidised by oxygen injected. The heat released from the oxidation of the char reheats the 'sand' over 980 °C. The hot 'sand' is then sent to the gasifier to provide heat required by gasification reactions. The ash from the char combustor and sand particles captured are sent to landfill after being cooled and moistened. The tar produced in the gasifier is reformed to CO and $H_2$ with the presence of catalyst in a bubbling fluidized bed reactor. The syngas generated in the biomass gasifier goes through a cooling and clean-up process to remove $CO_2$ and $H_2S$. During this process, the tar is reformed in an isothermal fluidized bed reactor and the catalyst is regenerated. The cleaned syngas is then converted to alcohols in a fixed bed reactor. The produced alcohol stream is depressurised in preparation of dehydration and separation afterwards. The evolved syngas in alcohol stream is recycled to the Gas Cleanup & Conditioning section. Finally, the alcohol mix is separated to methanol, ethanol and other higher molecular weight alcohols. The heat required for the gasifier and reformer operations and electricity for internal power requirements is provided by a conventional steam cycle. The steam cycle produces steam by recovering heat from the hot process streams throughout the plant.

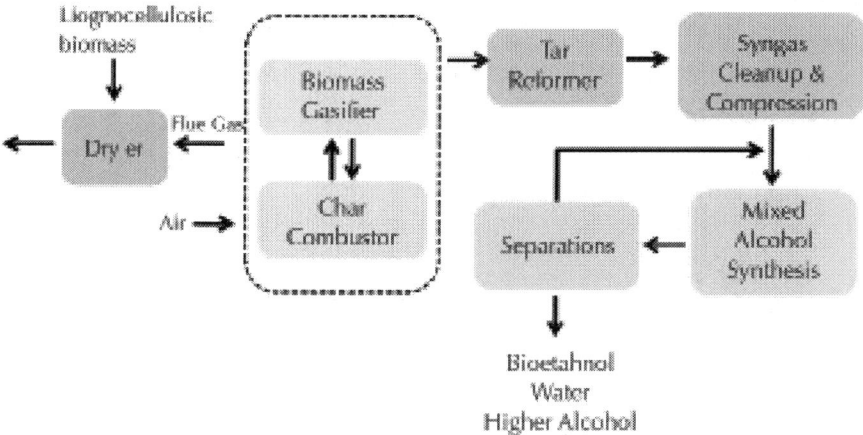

**Figure 3:** The schematic of a thermo-chemical cellulosic ethanol production process [57].

To compare these two approaches (biochemical *vs.* thermo-chemical) for producing bioethanol from economic point of view, process simulation and economic analysis are usually performed to calculate the minimum ethanol selling price (MESP) calculated from the discounted cash flow method. The MESP is defined as the selling price of bioethanol that makes the net present value of the biomass to bioethanol process equal to zero with a certain discounted cash flow rate with in a return period over the life of the plant [37]. In other words, it refers to the ethanol price at the break-even point which means annual costs and income are equal at this price. Several studies suggested that the estimated prices for 2G bioethanol produced biochemically is in the range of 2.16 to 4.44 USD $/gallon, depending on the type of biomass feedstock, technologies applied and the reference year based on [37,58-61]. On the other hand, NREL (National Renewable Energy Laboratory) reported a relatively low MESP for bioethanol produced thermo-chemically as 1.07 USD $/gallon. Nevertheless, raw materials cost (including biomass feedstock and catalyst or enzyme) is the main contributor to the MESP. For example, the cost of corn stover accounts for 40% and 43% of the MESP for bioethanol biochemically and thermo-chemically produced respectively [37, 57].

From environmentally point of view, a comparative LCA study showed that biochemical approach offers a slightly better performance on greenhouse gas emission and fossil fuel consumption impact categories, but the thermo-chemical pathway has significantly less water consumption [62].

# Butanol

Butanol is another attention attracted alternative fuel to gasoline besides ethanol because of its properties with respect to gasoline blending, distribution and refuelling, and end use in existing vehicles. For instance, butanol has relatively high energy content which is 30% higher than ethanol and is closer to gasoline. Additionally, butanol has low vapor pressure, low sensitivity to water and it is less volatile, and less flammable when compared with other liquid fuels [63]. Therefore butanol can be handled conventionally in the existing petroleum infrastructure, including transport *via* pipeline. It also can be blended, at any ratio, with either gasoline or diesel fuel at existing refineries, thus avoiding the capital investment associated with plant revamps and the need for major operational, *etc.*

Similarly to bioethanol, butanol can be biochemically produced from both agricultural crops and lignocellulosic biomass using *Clostridium acetobutylicum* or *C. beijerinckii* to ferment lignocellulosic hydrolysate sugars (hexoses and pentoses) to butanol. Traditionally, sugar-rich agricultural crops such as corn, cane molasses and whey permeate have been successfully used as feedstocks in the commercial production of butanol for decades. However, the cost for these food crops rises significantly nowadays; therefore, lignocellulosic biomass becomes more popular as substrates for butanol production. In similar ways of producing bioethanol, pre-treatments are required prior to enzymatic hydrolysis (using cellulase and cellobiose). However, one of technology challenges is the inhibition caused by by-products in pre-treatments such as furfural, HMF, acetic acid, and ferulic acid generated in dilute acid pre-treatments *etc*. Among these by-products, ferulic and ρ-coumaric acids were found can significantly inhibit fermentation but furfural and HMW were surprisingly stimulating to the cell culture [64].

The resulted lignocellulosic hydrolysate is then fermented by microorganisms *via* Acetone-butanol-ethanol (ABE) fermentation (Figure 4). The main challenge in the ABE fermentation is the product butanol itself is toxic to the fermenting microorganisms. In order to overcome this drawback, focused research efforts are to (1) improve the fermentation strategies to minimise the level of inhibitors accumulated such as simultaneously removing butanol and (2) to develop or genetically improve butanol – producing cultures.

However, biobutanol has several potential shortcomings. It is more toxic to humans and animals in the short term than ethanol or gasoline (although some components of gasoline, such as benzene, are more toxic and/or carcinogenic). And it is not clear whether butanol will degrade the materials commonly used in automobiles that can come into contact with motor fuels; building evidence suggests that it will not cause problems, but there has been no definitive testing on the wide range of potentially affected polymers and metals [65].

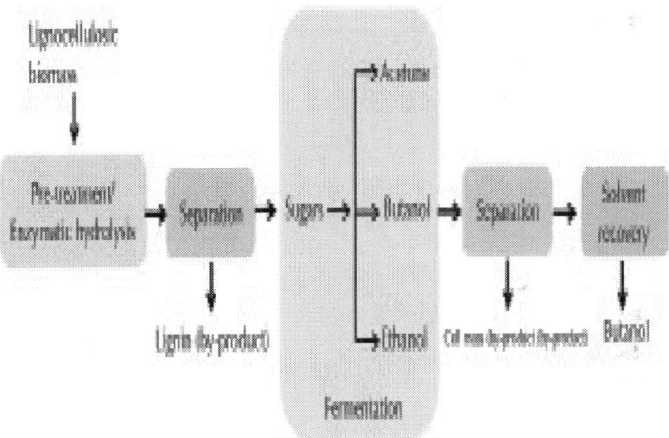

**Figure 4:** Phases of ABE fermentation for producing butanol.

Additionally, butanol is reported cannot deliver a better economic feasibility and a more sustainable environmental performance when compared with bioethanol under the current level of technology [66]. The relatively low yield of solvents out of glucose (mixture of acetone, ethanol and butanol), which is in the range of 33% - 45% (wt), is the main cause for the high cost of butanol. This economic study argued that butanol perhaps can be sold as chemicals rather than transport fuel unless the technology would be improved to make butanol production economically competitive with bioethanol.

# Biodiesel

Biodiesel refers to a liquid fuel alternative to petroleum diesel which can be used alone or blended with petroleum diesel. Similarly to bioethanol blends, blends of 20% biodiesel (B20) or lower can be used in diesel equipment without or with only minor modifications. Biodiesel can be produced from animal fat or oil from plants such as soybean and *Jatropha*, or from microalgae and fungi.

## *Biodiesel from Vegetable Oil*

Conventionally, the biodiesel is produced from vegetable oil with the presence of alcohol/alkaline/acid catalyst. This process is known as transesterificaiton or alcoholysis as shown in Figure 5 [67].

The vegetable oil is converted to esters and glycerol by reacting with an alcohol which can be ethanol, methanol or butanol. During this reaction, catalysts (e.g. alkalis, acids or enzymes) are required to improve the reaction rate and yield. Alkalis including NaOH, KOH and carbonates *etc.* are usually used as catalyst when feedstock containing less than 4% fatty acids. Acids, which are normally used when feedstocks contain more than 4% free fatty acids, include sulfuric acid, hydrochloric acid and sulphonic acids *etc.* Lipase, an enzyme that catalyses the hydrolysis of fats, can be used as a biocatalyst [68].

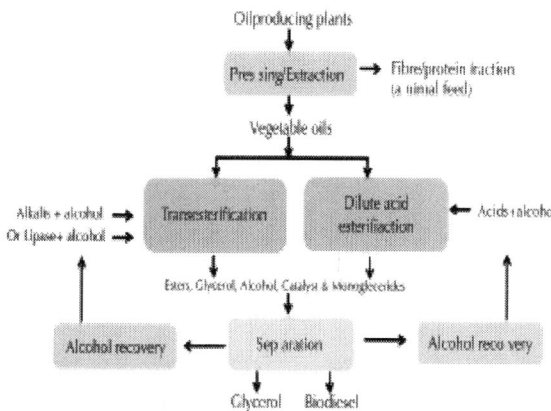

**Figure 5:** The schematic of biodiesel production [67].

A review by Ma and Hanna [69] summarized the parameters significantly influencing the rate of transesterificaiton reaction which include temperature, ratio of alcohol to oil, type of catalyst and catalyst concentration. The ester yield is increased by rising the transesterificaiton temperature; however, it will increase the risk of forming methanol bubbles when the temperature is close to methanol's boiling point. The ratio of alcohol to oil depends on the type of catalyst used which is approximately 6:1 for alkali catalyst and 30:1 for acid catalyst [70]. Enzyme used as a catalyst is becoming more attractive nowadays because it tolerates free fatty acid and water contents in the oil to avoid soap formation and thus results in an easier purification of biodiesel and glycerol [68]. However, the relatively high price of enzyme catalyst makes its utilization in the commercial production of biodiesel challenging.

Nowadays, 90% of U.S. biodiesel is made from soybean oil. The price relationship between vegetable oils and petroleum diesel is key influential factor to the profitability of biodiesel industry. Because of the increasing price of vegetable oils, biodiesel industry is suffering uncomfortable situations [71]. As a result, alternative non-food feedstocks and the associated technologies are becoming the focused research in biodiesel area.

*Jatropha curcas* is an agro-forestry crop growing in tropical and sub-tropical countries, such as India, Sahara Africa, South East Asia and China. This crop grows rapidly and takes 2-3 years to reach maturity with economic yields [72]. Lu *et al.* reported a higher than 98% biodiesel yield by a pre-esterification using solid acid followed by a transesterificaiton using KOH [73]. A high yield of 98% (wt) is also reported by Shah *et al.* [74] which is obtained from *Jatropha* oil using *Pseudomonas cepacia* lipase. Kumari *et al.* [75] also documented a relatively high yield of 94% (wt) biodiesel yield from *Jatropha* oil using lipase from *Enterobacter aerogene*. They also reported negligible loss in lipase activity even after repeated use for several cycles.

## Biodiesel from Microalgae

Due to biodiesel produced from oil crops, waste cooking oil and animal fat cannot meet the high demand for renewable transport fuels, another biomass feedstock microalgae becomes attractive. This is because (1) microalgae are sunlight-driven cells, (2) grow rapidly with biomass double time of 24 hours, (3) require less high quality land used compared to other feedstock, (4) many are exceedingly rich in oil and (5) biodiesel produced from microalgae is 'carbon neutral' [76] (see Figure 6). However, several challenges need to be tackled in order to produce biodiesel from microalgae commercially. Scott *et al.* [77] provides a comprehensive review discussing these challenges and potential tackles.

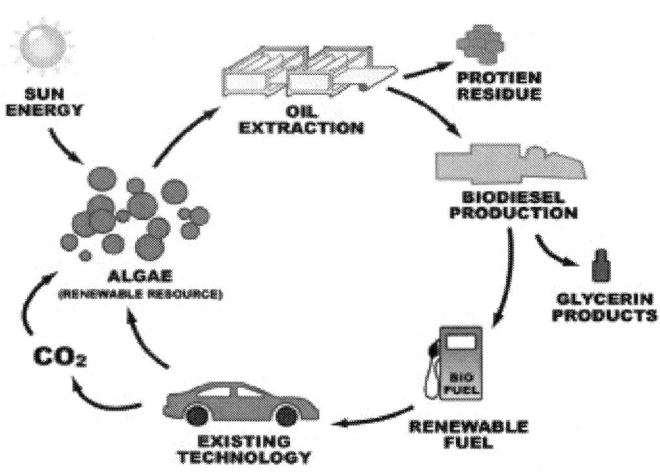

**Figure 6:** Life cycle of biodiesel produced from microalgae.

There are estimated 300 000 species in algal strain. After screening, typical species including*Botryococcus braunii, Nannochloropsis sp., Neochloris oleoabundans, Nitzschia sp.,* and*Schizochytrium sp.* have up to 77% (dry wt) oil content [76]. Microalgal biomass is produced with the presence of light, fed carbon dioxide and essential inorganic elements including nitrogen (N), phosphorus (P), iron and in some cases silicon. Biomass is then harvested and extracted to obtain oil for biodiesel production using transesterificaiton with methanol. Nutrients and spent biomass are recycled in the downstream process.

Factors involved in these phases are all important to be considered and optimized to maximize the biomass yield and minimize the production cost. First of all, the light level needs to be manipulated to deliver an optimal light to all of the algae cells within the culture.

The excess light level not only can results in less efficient use of absorbed light energy but also can cause biochemical damage to the photosynthetic machinery [77]. Secondly, though minimal nutrients requirement can be estimated according to the approximate molecular formula of microalgae which is $CO_{0.48} H_{1.83} N_{0.11} P_{0.01}$ [78], nutrients such as phosphorous must be supplied in excess. In order to minimise the nutrient cost, sea water supplement with commercial nitrate and phosphate fertilisers can be used for growing microalgae [76]. Thirdly, the choice of facility (open raceway ponds or closed photobioreactor) is important since the scale-up of biomass production is largely depending on the surface area rather than volume because light only penetrate a few centimeters [77]. The former raceway pond is an open-top close loop recirculation channel with a typical depth of 0.3 m. It is relatively cheap to build and has been operated with extensive experience for decades. However, the drawbacks for this type of facility are (1) it is difficult to avoid microbial contamination, (2) it requires for extensive areas of land for ht raceways and substantial cost regarding harvesting, and (3) it has poorly mixed therefore has optically dark zone [76, 77]. The photobioreactor a tubular reactors consists of an array of glass or plastic transparent tubes. It requires a large amount of energy for pumping and compressing air for sparging culture [77].

The biomass broth from production phase is harvested and processed to remove water and residual nutrients which are recycled. The concentrated biomass paste is then extracted to obtain oil and lipids using water and extraction solvent (e.g. hexane) [79]. It is difficult to release lipids from microalgae intracellular location using an energy-efficient way because of the large amount of solvent required. Also it is key to avoid significant contamination by other cellular components such as DNA [77].

The efforts in academic research and industrial commercialization of biodiesel production from microalgae include: (1) integration of production process such as energy integration, water and nutrient recycling; (2) improvement of microalgae biology *via* genetic and metabolic engineering such as enhancing their photosynthetic efficiency, increasing biomass yield and oil content and improving temperature tolerance to reduce cost associated with cooling; (3) improving photobioreactors regarding their capacity and operational ability [76].

# Biogas from Anaerobic Digestion

Anaerobic digestion (AD) has been used to treat biodegradable solid waste such as MSW, industrial waste and sewage sludge over decades. Biogas containing methane and carbon dioxide is the main product form AD digester. Generally, biogas is collected in the gas tank and they can be directly exported to national gas grid or sent to combustion in the CHP system to generate electricity (with a yield in the range of 0.7 – 2.0 kwh/m$^3$ biogas) and heat.

AD process is a dynamic complex system involving microbiological, biochemical, and physical-chemical processes though which the biodegradable waste are turned into biogas. Among biological waste treatment methods, AD has been identified as the most environmentally sustainable option for treating biowaste since it offers a unique technology which enables not only diverting biodegradable from landfill but also producing bio-energy and potential by-products such as a beneficial soil conditioner [80].

AD systems generally have four classifications [80]:

Mesophilic (30 - 40 ˚C) or thermophilic (50 - 65˚C) according to temperature

Wet digestion (< 15% dry solid) or dry digestion ( between 20% - 40% dry solid) according to the solid content in feedstock

Single step (one vessel) or multiple step digestions (normally two-step digestion i.e. hydrolysis and methanogenesis)

Batch digestion (loading feedstock in the beginning and remove products at the end of process) or continuous digestion (loading feedstock and withdraw products continuously)

Generally, five trophic groups are considered to be relevant to the process such as hydrolysing bacteria, acidogenic bacteria, acetogenic bacteria, aceticlastic and hydrogentrophic methanogens. They are involved in a series of digestion steps which are described as following and in Figure 7 [81] :

Carbohydrates, lipids, proteins *etc.* are broken down through hydrolysis to sugars, long – chain fatty acids and amino acids by extracellular enzymes released by hydrolytic bacteria;

Then these molecules are converted into volatile fatty acids, alcohols, $CO_2$ and $H_2$ in acidogenesis step;

These molecules are then further converted by acetogenic bacteria mainly into acetic acid, $H_2$ and $CO_2$;

Finally, all these intermediate products are turned into $CH_4$, $CO_2$ and water in the last step where methanogenic bacteria are involved. Three biochemical pathways are used by methanogens to produce methane gas:

- Acetotrophic methanogenesis: $4\ CH_3COOH ->4\ CO_2 + 4\ CH_4$
- Hydrogenotrophic methanogenesis: $CO_2 + 4\ H_2-> CH_4 + 2\ H_2O$
- Methylotrophic methanogenesis: $4\ CH_3OH + 6\ H_2 -> 3\ CH_4 + 2\ H_2O$

Due to for different substances, biological consortia and digestion conditions, the overall biogas yield and methane content will vary. Typically, the methane content of biogas is in the range of 40-70 % (v/v) [82].

Several key factors influence the Ad performance. They include pH, temperature, organic loading rate (OLR), the ratio of inoculum to substance (I/S) and the presence of inhibitory substances. Generally, mesophilic AD (35 - 37 °C) is more preferred than thermophilic AD (50 - 60°C) since the latter one offers less methane yield and it is more sensitive to environment change [81]. The pH range suggested for AD process is in the range of 6.8 -7.2 [80]. In addition, anaerobic digestion requires attention to the loading of nutrients for bacteria including carbon and nitrogen. The proper ratio of these two components (C/N) depends on the digestibility of the carbon and nitrogen sources between 20: 1 and 30:1. Other nutrients such as S, Mg, K, P, Ca, Fe, Zn, Al, Ni, Co, Cu and vitamin B12 are necessary [80]. However, these components are generally contained in the Organic Fraction of MSW (OFMSW) while they are added in the laboratory scale AD systems.

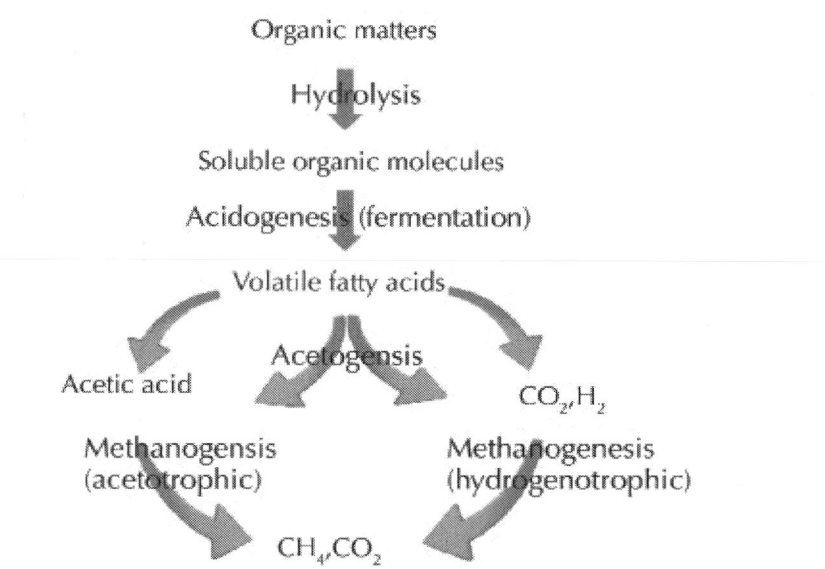

**Figure 7:** Anaerobic digestion biochemical conversion pathways.

Regarding the AD process operation, I/S ratio is considered as one the most important parameter. It is suggested to be approx 1 by Raposo *et al.*[83] who found that biogas production was inversely proportional to the I/S ratio in the range of 1 to 3. two stage AD is more preferred because it provides optimum environmental conditions for each bacteria group, offers accelerated digestion rates, better stability and thus increased methane yield [80]. Another process parameter is retention time which includes hydraulic retention time and solid retention time. The former refers to the mean time that any portion of liquid feed remains in a digestion system; the latter is defined as the mean time for any portion of solid feed or microbal biomass remains in the digester. These two retention times are the same in a single stage digester; while in a two stage digestion system the longer solid retention time is, the higher degradation rates and biogas yields are obtained [80]. In addition to the above process parameters discussed, the organics loading rate (OLR) is also critical which is measured as volatile solid (VS) or chemical oxygen demand (COD) of feed to a unit volume of digester per unit time [80]. The range for OLR is suggested in the range of 6 - 9.7 kg VS/day/m³ which is varied with the biodegradability of feedstock and AD systems [81].

Furthermore, the quality of OFMSW treated in biogas plants is also crucial for balanced performance of the biogas process, the technical feasibility of process and the use of residual/effluent as agricultural soil conditioner. Therefore, the costs associated with waste collection, sorting and pre-treatment should be considered [84].

Currently, most of MSW in the U.S. are sent to composting as an alternative to landfill. This is because it is more difficult to treat OFMSW than treating wastewater or manure. In addition, the AD of OFMSW requires a large amount of investment and technological experience as well as a higher capital and operating cost than compositing and landfilling [82]. The relatively low gate fees for landfill in the U.S. and relatively low energy prices make AD difficult to be commercialized in the U.S compared to those in Europe [82]. However, in the UK, there is currently very little waste treatment using AD apart from the use of AD to treat sewage sludge and wastewaters [85].

However, LCA studies have shown that AD of MSW reduces environmental impacts and is more cost - effective (in Europe) on a whole system basis than composting or landfilling options [86, 87].

# COMMODITY CHEMICALS AND MATERIALS

Today, only a small numbers of chemicals are produced from lignocellulosic feedstocks via fermentation. Much less attention has been given to biomass as a feedstock for organic chemicals, while there has been a strong political and technical focus on using biomass to produce transportation fuels. However, replacement of petroleum-derived chemicals with those from biomass will play a key role in sustaining the growth of the chemical industry [88]. One way to replace petroleum is through biological conversion of lignocellulosic resources into products now derived from petroleum. The current developments especially in fermentation technologies, membrane technologies and genetic manipulation open new possibilities for the biotechnological production of market relevant chemicals from renewable resources [5].

In lignocellulosic feedstocks biorefinery processes, the sugars or some of the fermentation products can be chemically converted into a variety of chemicals, which could be used to form biological materials, such as protein polymers, xantham gum, and polyhydroxybutyrate. The lignin as remaining fraction from lignocellulosic feedstocks, could be converted through hydrogenation processes into materials, such as phenols, aromatics, and olefins, or simply burned as a boiler fuel for cost efficiency of the overall process. Currently, conventional chemicals include acetaldehyde, acetic acid, acetone, n-butanol, ethylene, and isopropanol can simply be derived from LCF. Appropriate organisms could then convert the sugars into the desirable products and co-products for this process. The advantage to such products is that the market is already established, and minimal effort is required to integrate these products into existing markets. However, co-product markets might be limited, and caution must be taken in considering their impact on overall economics, especially for large-scale implementation. A sequence of processes comparable to those employed for cellulosic ethanol production would be used to pre-treat the lignocellulosic biomass to open its structure for the weight of the feedstock. Therefore, lignocellulosic biomass might be expected as the low cost of raw materials could be converted to a variety of commodity chemicals and materials [20].

# Present Promising Commodity Chemicals and Materials from LCF Biorefinery

## Lactic Acid

Lactic acid represents a chemical with a small world market, and the market for traditional applications of lactic acid is estimated to be growing at about 3–5% annually. New products based on lactic acid may increase the world market share significantly, which includes the use of derivatives such as ethyl esters to replace hazardous solvents like chlorinated hydrocarbon solvents in certain industrial applications. In theory, one mole of glucose results in almost two moles of lactic acid. The recovery process for lactic acid is much more sophisticated than that of the ethanol fermentations, involving various precipitations, chromatographic and distillation steps [5].

Lactic acid can be converted to methyl lactate, lactide, and polylactic acid (PLA) by fermentation [89]. The PLA is a biodegradable polymer used as environmentally friendly biodegradable plastic, which can be the replacement for polyethylene terephthalates (PETs) [90]. Recently, attempts have been made to produce PLA homopolymer and its copolymer by direct fermentation by metabolically engineered [91], shows a great potential for utilizing lignocellulosic feedstock for the key biodegradable polymers. Efforts are also under way to develop efficient processes for converting biologically produced lactic and hydroxypropionic acids to methacrylic and acrylic acids [88].

Lactic acid can be produced either chemically or by microbial fermentation. A major disadvantage for chemical synthesis is the racemic mixture of lactic acid. Microbial fermentation offers both utilization of renewable carbohydrates and production of pure L- or D-lactic acid depending on the strain selected. Currently, most of lactic acid production is produced mainly from corn starch. However, the use of lignocellulosic feedstock for lactic acid production appears to be more attractive because they do not impact the food chain for humans. But the process for converting lignocellulosic feedstock into lactic acid is not cost efficient due to the high cost of cellulase enzymes involved in cellulose hydrolysis [92, 93]. In addition, the main bottleneck during the hydrolysis of lignocellulosic feedstock by cellulases is the inhibition on cellulase by glucose and cellobiose, which remarkably slows down the rate of lignocellulosic feedstock hydrolysis [94]. Economic improvements on the process are mainly focused on increasing the lactic acid tolerance, reducing the requirements for complex and cost intensive growth supplements and products recovery [95].

## Acetone–Butanol–Ethanol (ABE)

An acetone – butanol – ethanol blend (in a ratio of 3-6-1) may serve as an excellent car fuel, which can be easily mixed not only with petrol but also with diesel. ABE as a fuel additive has the advantage of a similar heat of combustion to hydrocarbons, and perfect miscibility with hydrocarbons, even when water is present. The fermentative production of ABE used to be the second largest industrial fermentation after ethanol production [5]. Product inhibition caused principally by butanol is the main problem that hindering commercial development of the fermentation process. One way to overcome this inhibition problem would be to couple the fermentation process to a continuous product removal technique, so that inhibitory product concentrations are never reached. However, even with continuous product removal, product formation in these systems does not proceed indefinitely, because of the inhibition caused by the accumulation of mineral salts in the reactor [96]. Due to the shortage of raw materials, namely corn and molasses, and to decreasing prices of oil, ABE fermentation is not profitable when compared to the production of these solvents from petroleum. During the 1950s and 1960s, ABE fermentation was replaced by petroleum chemical plants.

Currently, the production of mixtures of acetone, butanol and ethanol (ABE) by sugars derived from lignocellulosic feedstocks continues to receive attention because of its potential commercial significance. The traditional fermentative production of acetone–butanol– ethanol is batch anaerobic bacteria fermentation with Clostridia. The substrate consists of molasses, and phosphate and nitrogen sources. Instead of molasses other sugar sources like sugar from lignocellulosic feedstock can also serve as a raw material for fermentation [97].

# Xylan

As one of main polysaccharides in lignocellulosic biomass, xylan has a variety of applications in our everyday life and affects our well-being. For example, (1) xylans are important functional ingredients in baked products [98]; (2) xylans can be potentially used for producing hydrogels as biodegradable coatings and also encapsulation matrices in many industrial applications; (3) xyl, the main constituent from xylans, can be converted to xylitol which is used as a natural food sweetener and a sugar substitute [99]; (4) xylans can be used for clarification of juices and improvement in the consistency of beer [100]; (5) xylans are also important for livestock industry as they are critical factors for silage digestibility; (6) xylans are major constituents in non-nutritional animal feed [101]; (7) xylans can be converted to sugars and then further to fuels and chemicals; (8) enzymes that degrade xylan can facilitate paper pulping and biobleaching of pulp [100].

Xylans, the main component in hemicellulose, are heteropolysaccharides with homopolymeric backbone chains of 1,4 linked β-d-xylopyranose units. In addition to xylose, xylans may also contain arabinose, glucuronic acid or its 4-O-methyl ether, acetic, ferulic, and p-coumaric acids. Xylans can be categorized as linear homoxylan, arabinoxylan, glucuronoxylan, and glucuronoarabinoxylan. Depends on the different sources of xylan (i.e. soft- and hard- wood, grasses, and cereals), the composition of xylans differs [100].

Hemicellulose can be derived via chemical treatment or enzymatic hydrolysis. As discussed in Section 2.1.1, several pre-treatments listed in Table 1 are available to fractionate, solubilize and hydrolyze and separate hemicellulose from cellulose and lignin components. Generally, hemicelluloses are solublized by either high temperature and short residence time (270°C, 1 min) or lower temperature and longer residence time (190 °C, 10 min) [102]. However, some of chemical treatment result in hemicellulose degradation by-products such as furfural and 5-hydroxymethyl furfural (HMF) which are inhibitors for microorganisms involved in downstream fermentation if applicable.

Biodegradation of xylan requires enzymes including endo-$\beta$-1,4-xylanase, $\beta$-xylosidase, and several accessory enzymes, such as $\alpha$-L-arabinofuranosidase, $\alpha$-glucuronidase, acetylxylan esterase, ferulic acid esterase, and $p$-coumaric acid esterase, which are necessary for hydrolyzing various substituted xylans. The endo-xylanase attacks the main chains of xylans while $\beta$-xylosidase breaks xylooligosaccharides to monomeric sugar xylose. The $\alpha$-arabinofuranosidase and $\alpha$-glucuronidase remove the arabinose and 4-$O$-methyl glucuronic acid substituents from the xylan backbone [100]. The esterases hydrolyze the ester linkages between xylose units of the xylan and acetic acid (acetylxylan esterase) or between arabinose side chain residues and phenolic acids, for example ferulic acid (ferulic acid esterase) and $p$-coumaric acid ($p$-coumaric acid esterase) [100].

Hemicellulose hydrolysates from lignocellulosic biomass either obtained by chemical treatment or enzymatic hydrolysis are attractive feedstock for producing bioethanol, 2,3-butanediol or xylitol. Other value added products from hemicellulose hydrolysate include (1) ferulic acid, and (2) lactic acid which can be used in the food, pharmaceutical, and cosmetic industries [100].

# Other Main Chemicals and Materials from Lignocellulosic Feedstock

Acetic acid, at present, most demand of the commercial acetic acid is met synthetically. The production involves fermentation by a species of Acetobacter, which converts ethanol to acetic acid with a small final concentrations percentage (4–6%), using almost exclusively for vinegar production. In commercial practice, the actual yield roughly 75–80% of the theoretical yield [5].

Ferulic acid, as a precursor for numerous aromatic chemicals used in the chemistry industry, can be produced from lignocellulosic feedstock [88].

Levulinic Acid, Formic Acid and Furfural, their biorefinery process usually involves the use of dilute acid as a catalyst but it differs from other dilute-acid lignocellulosic-fractionating processes in that free monomer sugars are not the product. Instead, these monosaccharides are converted into the platform chemicals levulinic acid and furfural as the final products by multiple acid-catalysed reactions [103].

# Opportunities and Challenges

New products from lignocellulosic feedstock including new adhesives, biodegradable plastics, degradable surfactants, and various plastics and polymers could also be derived through the unique biotechnologies. The products with desirable properties that are not easily matched by petrochemical processing are particularly promising targets. Therefore, less price pressure would exist initially for such new products. However, to have a substantial impact on petroleum consumption, it is necessary to ensure that large markets have to be eventually resulted [20].

Even today, the potential of microorganisms for the production of bulk chemicals is far from being fully exploited. The cost of feedstocks still remains one of the crucial points if biotechnological processes are to succeed. The transition of industrial chemical production from petrochemical to biomass feedstock faces real hurdles. Biorefinery processes do not require the high pressures and temperatures compared with most non-biological chemical processes, thus have the potential to reduce costs. However, current non-biological chemical processes (often continuous, and well integrated) for production of commodity chemicals have become highly efficient by evolved through considerable investment. Therefore biorefinery processes for production of commodity chemicals must rapidly approach similar levels of efficiency and productivity. Nevertheless, available technologies, economic opportunities, and environmental imperatives make the use of lignocellulosic feedstock and biorefinery for industrial chemical production not only feasible but highly attractive from multiple perspectives [88].

Simple criteria have been devised to allow rapid screening of potential chemicals and materials from lignocellulosic feedstock for their economic merit. We now need to identify products that have economic potential and improve the technology to a point where these technologies can be applied in a cost-effective way [20].

# FRACTIONATION OF LIGNOCELLULOSIC FEEDSTOCK

## Definition

Conversion of lignocellulosic materials to higher value products requires fractionation of the material into its components: lignin, cellulose, and hemicellulose, which convert to fuels, and chemicals for the production of most of our synthetic plastics, fibres, and rubbers is technically feasible. Liquefaction of LCF might serve as feedstocks for cracking to chemicals in the similar way that crude oil is presently used. Currently commercial products of LCF fractionation include levulinic acid, xylitol, and alcohols [104]. The ultimate goal of LCF fractionation is the efficient conversion of lignocellulose materials into multiple streams that contain value-added compounds in concentrations that make purification, utilization, and/or recovery economically feasible [15].

Fractionation of LCF is being developed as a means to improve the overall biomass utilization. Hemicellulose when separated from the LCF may find broader use for chemicals, fuel, and food application. The lignin separated in the process can be used as a fuel [105]. Unlike the lignin generated from pulping process, lignin fractionated from biomass by our approach is relatively clean, free of sulphur or sodium.

Fractionation of lignocellulosic materials is very difficult to accomplish efficiently, because of their complex composition and structure [106, 107]. However, fractionation of lignocellulosic materials is essential for some important applications, for example, paper-making, and in their conversion into basic chemical feedstocks or liquid fuels.

Figure 8 shows that fractionation of lignocellulosic biomass into its three major components, cellulose, hemicelluloses and lignin. It has been proposed as the first step of LCF refining to high value-added products [108]. Achieving high fractionation yields and maintaining the integrity of the macromolecular fractionation products are of major importance regarding the effectiveness of the whole refining process [109].

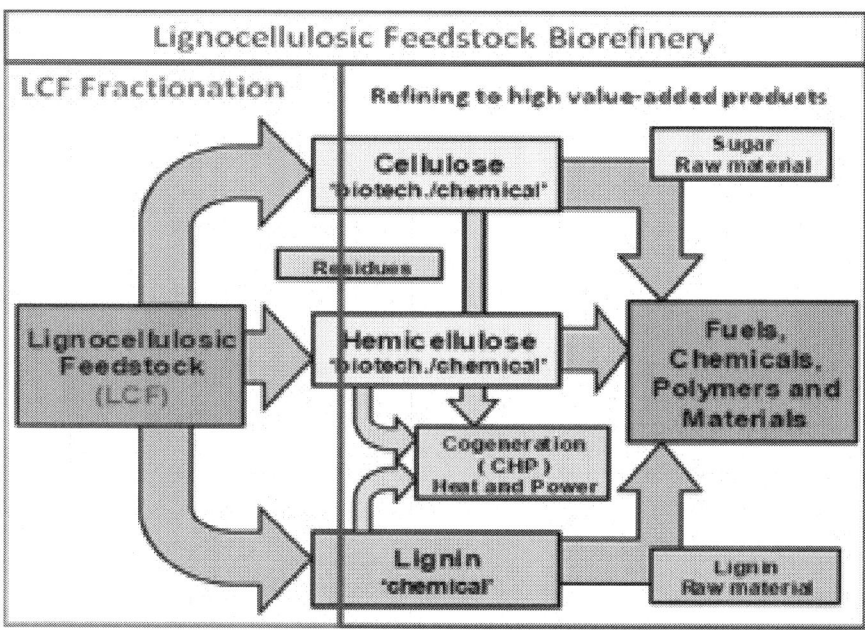

**Figure 8:** Lignocellulosic Feedstock Biorefinery [110].

# Organosolv Fractionation

The organosolv process is a unique and promising LCF fractionation. Using organosolv, lignocellulosic biomass can be converted into cellulosic fibres, hemicellulose sugars and low molecular weight lignin fractions in one-step fractionation [111-113]. Organosolv fractionation is the process to using organic solvents or their aqueous solutions to remove or decompose the network of lignin from lignocellulosic feedstocks with varying simultaneous hemicellulose solubilisation [114]. In this process, an organic or aqueous organic solvent mixture with or without an acid or alkali catalysts is used to dissolve the lignin and part of the hemicellulose, leaving reactive cellulose in the solid phase [106, 115-117]. Usually, the presence of catalyst can increase the solubilisation of hemicellulose and the digestibility of substrate is also further enhanced [118]. Comparing to other chemical pre-treatments the main advantage of organosolv process is that relatively pure, low molecular weight lignin is recovered as a by-product [119]. Organic solvents are always easy to recover by distillation and recycled for fractionation; the chemical recovery in organosolv fractionation processes can separate lignin as a solid material and carbohydrates as syrup, both of which can be used as chemical feedstocks [112, 120, 121]. A variety of organic solvents have been used in the organosolv process such as ethanol, methanol, acetone, ethylene glycol, triethylene glycol, tetrahydrofurfuryl alcohol, glycerol, aqueous phenol, aqueous n-butanol, esters, ketones, organic acids, *etc* [117, 119, 122]. For economic reasons, among all possible solvents, the use of low-molecular-weight alcohols with lower boiling points such as ethanol and methanol has been favoured [123].

Organic solvents are costly and their use requires high-pressure equipment due to their high volatility. The applied solvents should be separated from the system is necessary because the residual solvents may be inhibitors to enzymatic hydrolysis and fermentation [106], and they should be recycled to reduce operational costs. Otherwise organic solvents are always expensive, so it should be recovered as much as possible, but this causes increase of energy consumption.

The organosolv fractionation seems more feasible for biorefinery of lignocellulosic biomass, as it considers the utilization of all the biomass components. However, there are inherent drawbacks to the organosolv fractionation. In order to avoid the re-precipitation of dissolved lignin, the fractionated solids have to be washed with organic solvent previous water washing, the cumbersome washing processes means more cost. In addition, organosolv fractionation must be performed under extremely tight and efficient control due to the volatility of organic solvents. No digester leaks can be tolerated because of inherent fire and explosion hazard [121]. Its successful commercialization will depend on the development of high-value co-products from lignin and hemicelluloses [124].

# Ionic Liquids Fractionation

The ionic liquids (ILs) is a group of promising green solvents for the efficient fractionation of lignocellulosic materials. This technology has been used for delignification of lignocellulosic materials in paper-making [125]. Moreover, by fractionating lignocelluloses with ionic liquids it is possible to extract cellulose cleanly, which establishes a platform for the development of cellulose composites and derivatives.

ILs are liquid salts exist at relatively low temperatures (often at room temperature), which typically composed of large organic cations and small inorganic anions. By adjusting the anion and the alkyl constituents of the cation, ILs' solvent properties can be varied. The solvent properties include chemical and thermal stability, non-flammability, low vapour pressures and a tendency to remain liquid in a wide range of temperatures [126]. ILs are called "green" solvents, as no toxic or explosive gases are formed.

Most ILs are nonflammable and recyclable solvents with very low volatility and high thermal stability. Carbohydrates and lignin can be simultaneously dissolved in ILs, and the intricate network of non-covalent interactions between biomass polymers of cellulose, hemicellulose, and lignin is effectively disrupted while minimizing formation of degradation products [127-129].

ILs can dissolve large amounts of cellulose at considerable mild conditions and feasibility of recovering nearly 100% of the used ILs to their initial purity makes them attractive [130]. ILs as cellulose solvents, comparing with regular volatile organic solvents of biodegradability, possesses several advantages including low toxicity, broad selection of anion and cation combinations, low hydrophobicity, low viscosity, enhanced electrochemical stability, thermal stability, high reaction rates, low volatility with potentially minimal environmental impact, and non-flammable property.

However, ILs fractionation using ionic liquids faces many challenges in putting these potential applications into industrial scale., for example, the high cost of ILs, regeneration requirement [16]. Their toxicity toward enzymes and microorganisms must also be established before ILs can be considered as a real option for LCF pre-treatment [129].

Other main challenges are the recovery of ionic liquids and the recovery of hemi-cellulose and lignin from the ionic liquids after extraction of cellulose [126].

# Liquid Hot Water (LHW) Fractionation

Liquid hot water fractionation does not employ any catalyst or chemicals. Pressure is utilized to maintain water in the liquid state at elevated temperatures (160–240 °C) and provoke alterations in the structure of the lignocelluloses [131-133]. LCF in LHW undergoes high temperature cooking in water with high pressure. LHW pre-treatment has been reported to have the potential to enhance cellulose digestibility, sugar extraction, and pentose recovery, with the advantage of producing hydrolysates containing little or no inhibitor of sugar fermentation [134].

Water is an abundant, non-toxic, environmentally benign and inexpensive solvent. LHW is the part range of sub-critical water that near its critical point (374 °C, 22.1 MPa), Sub-critical water (SCW) possesses marvellous properties which are very different from that of ambient liquid water [135-138]. In SCW, dielectric constant, surface tension, and viscosity decrease dramatically with increasing temperature, which enhances the solubility of organic compounds. Sub-critical water is more like non-polar organic solvent (similar with acetone), thus it can substitute for some of organic solvents, and become a clean medium for chemical reactions. SCW is a tunable reaction medium for conducting ionic/free radical reactions, and an effective medium for energy and mass transfer. The ionic product of SCW is larger by three orders of magnitude than that of ambient water, which means concentrations of hydrogen and hydroxide ions are much higher. Therefore, in addition to the increase in kinetic rates with temperature, both acid and base catalyses by water are enhanced in SCW, which can be a solvent or reactant participated in chemical reaction. And without any pollution, hydrolysis in SCW is an environment-friendly technology

The objective of the liquid hot water is to solubilise mainly the hemicellulose to make the cellulose more accessible and to avoid the formation of inhibitors. By keeping the pH between 4 and 7 the autocatalytic formation of fermentation inhibitors are avoided during the fractionation [34, 139, 140]. If catalytic degradation of sugars occurs it results in a series of reactions that are difficult to control and result in undesirable side products.

The slurry generated after pre-treatment can be filtered to obtain two fractions: one solid cellulose-enriched fraction and a liquid fraction rich in hemicellulose derived sugars [34]. Lignin is partially depolymerised and solubilised as well during hot water fractionation but complete delignification is not possible using hot water alone, because of the re-condensation of soluble components originating from lignin.

Water under high pressure can penetrate into the LCF, hydrate cellulose, and remove hemicellulose and part of lignin. The major advantages are no addition of chemicals and no requirement of corrosion-resistant materials for hydrolysis reactors in this process. Liquid hot water pre-treatments are attractive from no catalyst requirement and low-corrosion potential. Liquid hot water has the major advantage that the solubilised hemicellulose and lignin products are present in lower concentrations, due to higher water input and subsequently concentration of degradation products like furfural and the condensation and precipitation of lignin compounds is reduced. However, water demanding in the process and energetic requirement are higher and it is not developed at commercial scale [141].

# Combined Technology for LCF Fractionation

The efficiency of lignocelluloses utilization can be significantly improved by fractionation [40]. Fractionation of lignocellulosic materials may be achieved by various physical, chemical and biological methods. Combination of different methods may lead to more efficient fractionation processes of lignocellulosic materials [5].

The most promising combined technology for LCF fractionation is the combination of liquid hot water (LHW) with the assisted technologies, which usually are performed before or during the LHW fractionation, including steam explosion, $CO_2$ explosion, Ammonia fibre explosion (AFEX), acid or alkaline pre-treatment, High energy radiation pre-treatment, Wet oxidation and Ozonolysis etc.

## Combination with Steam Explosion

Steam explosion is the most widely employed physical-chemical pre-treatment for lignocellulosic biomass. It is a hydrothermal pre-treatment in which the biomass is subjected to pressurised steam for a period of time ranging from seconds to several minutes, and then the pressure is suddenly reduced and makes the materials undergo an explosive decompression. The treatment leads to the disruption of the structure of the material due to the rapid expansion of the water vaporized inside it. The temperatures involved are higher than, or close to, the glass transition temperature of hemicellulose, lignin and cellulose impregnated with water [142, 143], so that the internal cohesion of lignocelluloses is weakened and disaggregation and defibration of the material are facilitated. This pre-treatment combines mechanical forces and chemical effects due to the hydrolysis (auto-hydrolysis) of acetyl groups present in hemicelluloses.

Hydrolytic treatments of lignocellulosic biomass by saturated steam, with (un-catalyzed) and without (catalyzed) addition of small amounts of mineral acids, have been widely studied as a method to weaken the lignocellulosic structure and increase its chemical reactivity and enzyme accessibility [144,145].

Un-catalyzed steam-explosion is one of only a very limited number of cost-effective pre-treatment technologies that have been advanced to pilot scale demonstration and commercialized application [16]. Autohydrolysis takes place when high temperatures promote the formation of acetic acid from acetyl groups; furthermore, water can also act as an acid at high temperatures. The mechanical effects are caused because the pressure is suddenly reduced and fibres are separated owing to the explosive decompression. In combination with the partial hemicellulose hydrolysis and solubilisation, the lignin is redistributed and to some extent removed from the material [146]. Catalyzed steam-explosion is very similar to un-catalyzed steam-explosion on their action modes, except that some acidic chemicals (gases and liquids), primarily including $SO_2$, $H_2SO_4$, $CO_2$, oxalic acid, etc. are used as catalysts to impregnate the LCF prior to steam-explosion, to improve recovering both cellulose and hemicellulose fractions [147]. It is recognized as one of the most cost-effective pre-treatment processes

[148, 149]. Compared to un-catalyzed steam explosion, catalyzed steam-explosion has more complete hemicellulose removal leading to more increased enzymatic digestibility of LCF with less generation of inhibitory compounds [150]. A steam-explosion/separation process offers several attractive features when compared to the alternative hydrolysis and pulping processes. These include the potential for significantly lower environmental impact, lower capital investment, more potential for energy efficiency, less hazardous process chemicals and conditions [151]. Steam-explosion allows the recovery of all constitutive LCF components without the destructive degradation of any one component in favour of any other [152]. The process is generally followed by fractionation steps in order to separate the various components.

## Combination with $Co_2$ Explosion

Carbon dioxide explosion can also be used for lignocellulosic biomass pre-treatment. The method is based on the utilization of $CO_2$ as a supercritical fluid, which refers to a fluid that is in a gaseous form but is compressed at temperatures above its critical point to a liquid like density. Supercritical carbon dioxide has been used as an extraction solvent for non-extractive purposes, due to some advantages such as availability at relatively low cost, non-toxicity, non- flammability, easy recovery after extraction, and environmental friendly [153]. Besides a liquid-like solvating power, supercritical carbon dioxide displays gas-like mass transfer properties [154].

Supercritical pre-treatment conditions can effectively increase substrate digestibility by removing lignin. Addition of co-solvents such ethanol can improve delignification. Supercritical carbon dioxide has been mostly used as an extraction solvent but it is being considered for non-extractive purposes due to its many advantages [155]. $CO_2$ molecules are comparable in size to water and ammonia and they can penetrate in the same way the small pores of lignocelluloses. This mechanism is facilitated by high pressure. After $CO_2$ explosive, pressure released, disruption of cellulose and hemicellulose structure is observed and consequently accessible surface area of the substrate to enzymatic attack increases [141].

# Combination with Ammonia Fibre Explosion (AFEX)

Similar to steam explosion, AFEX is one of the alkaline physical-chemical pre-treatment processes. Here the biomass is exposed to liquid ammonia under high pressure for a period time, and then the pressure is suddenly released, resulting in a rapid expansion of the ammonia gas that causes swelling and physical disruption of LCF fibres and partial decrystallization of cellulose. This swift reduction of pressure opens up the structure of lignocellulosic biomass leading to increased digestibility of biomass.

One of the main advantages of AFEX pre-treatment is no formation of some types of inhibitory by-products, which are produced during the other pre-treatment methods, such as furans in steam explosion pre-treatment.

AFEX has been studied for decreasing cellulose crystallinity and disrupt lignin–carbohydrates linkages [156]. Ammonia recovery and recycle is feasible despite of its high volatility [157] but the associated complexity and costs of ammonia recovery may be significant regarding industrial scale using of the AFEX pre-treatment [34, 158].

There are some disadvantages in using the AFEX process compared to some other processes. AFEX simultaneously de-lignify and solubilize some hemicellulose while decrystallizing cellulose, but does not significantly solubilize hemicellulose as acid and acid-catalyzed steam-explosion pre-treatments [159-161]. The AFEX produces only a pre-treated solid fraction, while steam explosion produces a slurry that can be separated in a solid and a liquid fractions [15]. Furthermore, ammonia must be recycled after the pre-treatment to reduce the cost and protect the environment [106, 158].

# Combination with Acid or Alkaline Treatment

A way to improve the effect of LHW fractionation is to add an external acid or alkali, which can catalyze the solubilisation of the hemicellulose, reduce the optimal pre-treatment temperature and gives a better enzymatic hydrolysable substrate [162-164].

Acid pre-treatments can be performed with concentrated or diluted acid. However utilization of concentrated acid is less attractive for ethanol production due to the formation of inhibiting compounds, and high acid concentration (e.g. 30-70%) in the concentrated-acid process makes it extremely corrosive and dangerous [165, 166]. Diluted acid pre-treatment appears as more favourable method for industrial applications and have been studied for fractionation wide range of lignocellulosic feedstocks, including softwood, hardwood, herbaceous crops, agricultural residues, wastepaper, and municipal solid waste. It performed well on most biomass materials, mainly xylan, but also converting solubilised hemicellulose to fermentable sugars. Of all acid-based pre-treatment methods, sulphuric acid has been most extensively studied since it is inexpensive and effective. Organic acids such as fumaric or maleic acids are appearing as alternatives to pre-treat LCF for fractionation. Organic acids also can pre-treat lignocellulosic materials with high efficiency although fumaric acid was less effective than maleic acid. Furthermore, less amount of furfural was formed in the maleic and fumaric acid pre-treatments than with sulphuric acid [167]. Phosphoric acid, hydrochloric acid and nitric acid have also been tested [34].

Alkali pre-treatment refers to remove lignin and a part of the hemicellulose, by use of alkaline solutions such as NaOH and $Ca(OH)2$, and efficiently increase the accessibility of enzyme to the cellulose. Alkali pre-treatment can be used at room temperature and times ranging from seconds to days. It is reported to cause less sugar degradation than acid pre-treatment. It is basically a delignification process, in which a significant amount of hemicellulose is solubilised as well. Alkaline pre-treatment of lignocellulosic materials causes swelling, increasing the internal surface of cellulose and decreasing the degree of polymerization and crystallinity, which provokes lignin structure disruption, and separation of structural linkages between lignin and carbohydrates [117]. In general, alkaline pre-treatment is more effective on hardwood, herbaceous crops, and agricultural residues with low lignin content than on softwood with high lignin content [168]. Alkali pre-treatment was shown to be more effective on agricultural residues than on wood materials [169]. Addition of an oxidant agent (oxygen/$H_2O_2$) to alkaline pre-treatment (NaOH/ $Ca(OH)_2$) can favour lignin removal to improve the performance [170].

# Combination with Ammonia and Carbon Dioxide Solution

The aim of combination is to enhance alkaline or acidic intensity of liquid hot water by ammonia or carbon dioxide for lignocelluloses fractionation.

Ammonia is an extremely important widely used bulk chemical. The polarity of Ammonia molecules and their ability to form hydrogen bonds explains to some extent the high solubility of ammonia in water. In aqueous solution, ammonia acts as a base, acquiring hydrogen ions from $H_2O$ to yield ammonium and hydroxide ions.

$$NH_3 (aq) + H_2O (l) \approx NH_4{+}(aq) + OH{-}(aq) \qquad (1)$$

The production of hydroxide ions when ammonia dissolved in water gives aqueous solutions of ammonia the characteristics of alkaline properties.

Carbon dioxide can be considered as an ideal solvent for the treatment of natural products, because of the relatively low critical pressure (73.8 atm) and critical temperature (31.1 °C), it. In contrast with organic solvent, Super-critical carbon dioxide is non-toxic, non-flammable, non-corrosive, cheap and readily available in large quantities with high purity [171].

Carbon dioxide dissolves in water becomes acidic due to the formation and dissociation of carbonic acid:

$$CO_2 + H_2O \approx H_2CO_3 \approx H{+} + HCO_3 \qquad (2)$$

Over the temperature range 25-70 °C and pressure range 70-200 atm, the pH of solution ranged between 2.80 and 2.95, and increases with increasing temperature and decreases with increasing pressure [172]. It was shown that in the presence of water, supercritical $CO_2$ can efficiently improve the enzymatic digestibility of lignocellulosic materials [32].

## Combination with High Energy Radiation Treatment

Digestibility of lignocellulosic materials can be enhanced by the application of high energy radiation methods, such as microwave heating [173-175] and ultrasound [176, 177]. The treatments can cause hydrolysis of hemicellulose, and partial depolymerization of lignin, the increase of specific surface area, decrease of the degrees of polymerization and crystallinity of cellulose.

Microwave treatment is a physical-chemical process involving both thermal and non-thermal effects. Treatments can be carried out by immersing the biomass in dilute chemical reagents and exposing the slurry to microwave radiation for a period of time [178]. The treatment of ultrasound on lignocellulosic biomass have been used for extracting hemicelluloses, cellulose and lignin [179]. Some researchers have also shown that saccharification of cellulose is enhanced efficiently by ultrasonic pre-treatment [180]. The efficiency of ultrasound in the treatment of vegetal materials has been already proved [181]. The well-known benefits from ultrasounds, such as swelling of vegetal cells and fragmentation due to the cavitational effect associated to the ultrasonic treatment. Furthermore, mechanical impacts produced by the collapse of cavitation bubbles, give an important benefit of opening up the solid substrates surface for enzymatic hydrolysis [180].

However, the high energy radiation methods are usually energy-intensive and prohibitively expensive; appear to be strongly substrate-specific. The current estimation of overall cost from high energy radiation techniques looks too high, lack commercial appeal.

## Combination with Oxidative Treatment

### Wet Oxidation

Wet oxidation is an oxidative pre-treatment method which employs oxygen or air as catalyst, and can be operated at relatively low temperatures and short reaction times [182]. It is an exothermic process, therefore self-supporting with respect to heat while the reaction is started [183]. Wet oxidation of the hemicellulose fraction is a balance between solubilisation and degradation. Wet oxidation has been proven to be an efficient method for separating the cellulosic fraction from lignin and hemicellulose [184], and also been widely used for ethanol production followed by SSF [185]. Wet oxidation pre-treatment mainly causes the formation of acids from hydrolytic processes, as well as oxidative reactions. The hemicelluloses are extensively cleaved to monomer sugars, cellulose is partly degraded, and the lignins undergo both cleavage and oxidation in wet oxidation pre-treatment. Therefore lignin produced by wet oxidation cannot be used as a fuel [186]. In general, low formation of inhibitors and efficient removal of lignin can be achieved with wet oxidation pre-treatment.

### *Ozonolysis*

Ozone is a powerful oxidant that shows high delignification efficiency [106]. This method can effectively degrade lignin and part of hemicellulose. The pre-treatment is usually carried out at room temperature, and does not lead to inhibitory compounds [187]. It is usually performed at room temperature and normal pressure and does not lead to the formation of inhibitory compounds that can affect the subsequent hydrolysis and fermentation. However, ozonolysis might be expensive since a large amount of ozone is required, which can make the process economically unviable [106].

# OTHER BIOCONVERSION TECHNOLOGIES

## Landfill Gas (LFG) Production

As discussed in Section 2.4, anaerobic digester is a suitable waste treatment method to deal with wastewater, sewage sludge and animal mature since the high solid content of other types waste would challenge the anaerobic digester operation technologies. Currently most of biodegradable waste is sent to landfill where landfill gas (LFG) is generated.

Because the wastes sent to landfill include not only biodegradable components but also other hazard wastes, the LFG produced contains approx 40 - 60% methane, $CO_2$, and varying amounts of nitrogen, oxygen, water vapour, volatile organics (VOC), $H_2S$ and other contaminates (also known as non-methane organic compounds NMOCs). Some other inorganic contaminants, for example, heavy metals are found present in the LFG. Therefore, the direct release of the landfill gas to atmosphere will cause serious greenhouse gas emissions and pollutions. LFG produced from landfill site has to be monitored and managed appropriately. The general LFG managing options are: flaring (burn without energy recovery), boiler (produces heat), internal combustion (producing electricity), gas turbine (producing electricity), fuel cell (producing electricity), convert the methane to methyl alcohol, or sent to natural gas lines after cleaning process [188].

## Biopulping and Wood Utilization

Biopulping, also known as biological pulping, refers a type of industrial biotechnology using fungus to convert wood chips to paper pulp. This technology has the potential to improve the quality of paper pulp, reduce energy consumption and environmental impacts when compare with the traditional chemical pulping technologies [189].

The aim of pulping is to extract cellulose from plant material. The traditional approaches are mechanical and chemical pulping. The former method is generally accomplished by refining grinding or thermo-mechanical pulping. The latter way is to dissolve lignin from the cellulose and hemicellulose fibers via chemical treatment, such as kraft pulping in which wood chips are cooled in a solution containing sodium hydroxide and sodium sulfide [190]. These traditional pulping technologies have several drawbacks: (1) high energy demand; (2) low cellulose yield, especially from chemical pulping due to partial degradation of cellulose; (3) potential hazards chemicals emitted to the environment [189].

Lignin is a complex polymer which serves as a structural component of higher plants and is highly resistant towards chemical degradation [191]. White-rot and brown-rot fungi are two classifications of wood-rotting basidiomycetes. White-rot basidimycetes have been reported enable to, selectively or simultaneously with cellulose, degrade lignin in different types of wood [191, 192]. Brown-rot basidiomycetes, which grow mainly on softwood, can degrade wood polysaccharides but cause only a partial modification of lignin. Besides white- and brown-rot basidimycetes, some scomycetes so-called soft-rot fungi which can degrade wood under extreme environmental conditions such as high or low water potential that prohibit the activity of other fungi [191].

The fungal treatment process fits in a paper mill operation well. After wood is debarked, chipped and screened, wood chips are briefly steamed to reduce natural chip microorganisms, cooled with air, and inoculated with the biopulping fungus for 1 to 4 weeks prior to further processing. The biopulping has been indicated as a technology technologically feasible and economically beneficial [193].

This biological treatment of wood using fungi has also been studied and used as a pre-treatment approach prior to enzymatic hydrolysis for biofuel production [194-196]. However, more research are required to understand the mechanism of wood degradation, structural changes of wood cell wall caused by these wood decay fungus and to improve the treatment technologies [197, 198].

# CONCLUSIONS

The concept of 'biorefinery' has emerged since the potential of lignocellulosic based products substituting fossil fuel derived products has been discovered. Biorefienries may play a major role in tackling climate change by reducing the demand on fossil fuel energy and providing sustainable energy, chemicals and materials, potentially aiding energy security, and creating opportunities and market. This paper reviewed a wide range of such lignocellulosic derived products and current available biorefinery technologies. Some of these technologies have been or being close to the industrialization and others are still at the early stage of development. However, more research efforts are required to improve the technologies and integrate the biorefinery system in order to achieve the maximum outputs and to make biorefinery work at scale.

# REFERENCES

1. J. H. Clark, Green chemistry for the second generation biorefinery-sustainable chemical manufacturing based on biomassJournal of Chemical Technology & Biotechnology, 2007603609

2. M. Stöcker, Biofuels and Biomass-To-Liquid Fuels in the Biorefinery: Catalytic Conversion of Lignocellulosic Biomass using Porous Materials.Angewandte Chemie International Edition, 200892009211

3. P. Gullón, et alSelected Process Alternatives for Biomass Refining: A Review. The Open Agricultrure Journal 2010135144

4. S. S. Kelley, Lignocellulosic Biorefineries: Reality, Hype, or Something in Between?, in Materials, Chemicals, and Energy from Forest Biomass. 2007American Chemical Society. 3147

5. H. Danner, and R. Braun, ChemInform Abstract: Biotechnology for the Production of Commodity Chemicals from BiomassChemInform, 2000p. No-no.

6. J. N. Chheda, G. W. Huber, and J. A. Dumesic, Liquid-Phase Catalytic Processing of Biomass-Derived Oxygenated Hydrocarbons to Fuels and Chemicals.Angewandte Chemie International Edition, 200771647183

7. W. N. Rowlands, A. Masters, and T. Maschmeyer, The Biorefinery-Challenges, Opportunities, and an Australian Perspective.Bulletin of ScienceTechnology & Society, 2008149158

8. J. P. M. Sanders, B. Annevelink, and D. A. v. d. Hoeven, The development of biocommodities and the role of North West European ports in biomass chainsBiofuelsBioproducts and Biorefining, 2009395409

9. B. Kamm, Production of Platform Chemicals and Synthesis Gas from Biomass.Angewandte Chemie International Edition, 200750565058

10. B. Kamm, P. Schönicke, and M. Kamm, Biorefining of Green Biomass- Technical and Energetic Considerations. CLEAN- Soil, Air, Water, 20092730

11. L. Li, S. Lu, and V. Chiang, A. Genomic, and Molecular View of Wood Formation. Critical Reviews in Plant Sciences2006215233

12.   B. Kamm, et alBiorefinery Systems-An Overview, in Biorefineries-Industrial Processes and Products. 2008Wiley-VCH Verlag GmbH. 140

13.   S. Fernando, et alBiorefineries:? Current Status, Challenges, and Future Direction. Energy & Fuels, 200617271737

14.   B. Kamm, and M. Kamm, Biorefinery- Systems. Chemical and Biochemical Engineering Quarterly, 200416

15.   N. Mosier, et alFeatures of promising technologies for pretreatment of lignocellulosic biomass.Bioresource Technology2005673686

16.   Y. Zheng, Z. Pan, and R. Zhang, Overview of biomass pretreatment for cellulosic ethanol productionInternational Journal of Agricultural and Biological Engineering20095168

17.   P. Mckendry, Energy production from biomass (part 1): overview of biomass.Bioresource Technology20023746

18.   M. Kristian, and M. Hurme, Lignocellulosic biorefinery economic evaluation. Cellulose Chemistry and Technology, 2011443454

19.   W. Lynn, Biomass Energy Data Book: Edition 1.2006

20.   C. Wyman, and B. Goodman, Biotechnology for production of fuels, chemicals, and materials from biomassApplied Biochemistry and Biotechnology19934159

21.   A. García, et alBiorefining of lignocellulosic residues using ethanol organosolv processChemical Engineering Transactions2009911916

22.   D. Puppan, Environmental evaluation of biofuels. Periodica Polytechnica Ser. Soc. Man. Sci., 200295116

23.   Y. Lin, and S. Tanaka, Ethanol fermentation from biomass resources: current state and prospects.Applied Microbiology and Biotechnology2006627642

24.   R. B. Slade, Prospects for cellulosic ethanol supply-chains in Europe: a techno-economic and environmental assessment,in Centre for Process Systems Engineering and Centre for Environmental Policy. 2009Univiersity of London. 170

25.   T. Josefsson, H. Lennholm, and G. Gellerstedt, Steam Explosion of Aspen Wood. Characterisation of Reaction ProductsHolzforschung2002289297

26. I. Ballesteros, et alEffect of chip size on steam explosion pretreatment of softwood.Applied Biochemistry and Biotechnology200097110

27. E. Ruiz, et alEvaluation of steam explosion pre-treatment for enzymatic hydrolysis of sunflower stalks.Enzyme and Microbial Technology2008160166

28. M. Holtzapple, et alPretreatment of lignocellulosic municipal solid waste by ammonia fiber explosion (AFEX)Applied Biochemistry and Biotechnology1992521

29. H. Alizadeh, et alPretreatment of switchgrass by ammonia fiber explosion (AFEX).Applied Biochemistry and Biotechnology200511331141

30. C. Tengborg, et alComparison of $SO_2$ and $H_2SO_4$ impregnation of softwood prior to steam pretreatment on ethanol productionApplied Biochemistry and Biotechnology1998315

31. K. Öhgren, M. Galbe, and G. Zacchi, Optimization of Steam Pretreatment of SO2;-Impregnated Corn Stover for Fuel Ethanol Productionin Twenty-Sixth Symposium on Biotechnology for Fuels and Chemicals, B.H. Davison, et al., Editors. 2005Humana Press. 10551067

32. K. H. Kim, and J. Hong, C. O. Supercritical, pretreatment of lignocellulose enhances enzymatic cellulose hydrolysis. Bioresource Technology, 2001139144

33. Y. Zheng, H. M. Lin, and G. T. Tsao, Pretreatment for Cellulose Hydrolysis by Carbon Dioxide ExplosionBiotechnology Progress1998890896

34. 19861993Mosier, N., et al., Optimization of pH controlled liquid hot water pretreatment of corn stover. Bioresource Technology, 2005. 96(18): p. 1986-1993

35. M. Laser, et alA comparison of liquid hot water and steam pretreatments of sugar cane bagasse for bioconversion to ethanolBioresource Technology20023344

36. S. G. Allen, et alA Comparison between Hot Liquid Water and Steam Fractionation of Corn FiberIndustrial & Engineering Chemistry Research, 200129342941

37. A. Aden, M. Ruth, K. Ibsen, J. Jechura, K. Neeves, J. Sheehan, B. Wallace, L. Montague, A. Slayton, J. Lukas, Lignocellulosic

Biomass to Ethanol Process Design and Economics Utilizing Co-Current Dilute Acid Prehydrolysis and Enzymatic Hydrolysis for Corn Stover2002National Renewable Energy Laboratory (NREL). 95NREL/TP-510-32438.

38. R. Torget, M. E. Himmel, and K. Grohmann, Dilute sulfuric acid pretreatment of hardwood barkBioresource Technology, 1991239246

39. D. J. Schell, et alDilute-sulfuric acid pretreatment of corn stover in pilot-scale reactor- Investigation of yields, kinetics, and enzymatic digestibilities of solids. Applied Biochemistry and Biotechnology, 20036985

40. T. H. Kim, and Y. Y. Lee, Fractionation of corn stover by hot-water and aqueous ammonia treatmentBioresource Technology2006224232

41. C. Vaccarino, et alEffect of $SO_2$, NaOH and $Na_2CO_3$ pretreatments on the degradability and cellulase digestibility of grape marcBiological Wastes19877988

42. Chander KuhadR., et al., Fed batch enzymatic saccharification of newspaper cellulosics improves the sugar content in the hydrolysates and eventually the ethanol fermentation by Saccharomyces cerevisiaeBiomass and Bioenergy201011891194

43. R. Sierra, L. A. Garcia, and M. T. Holtzapple, Selectivity and delignification kinetics for oxidative short-term lime pretreatment of poplar wood. Part I: Constant-pressure.Biotechnology Progress2011976985

44. V. Chang, et alOxidative lime pretreatment of high-lignin biomassApplied Biochemistry and Biotechnology2001128

45. D. Pasquini, et alExtraction of lignin from sugar cane bagasse and Pinus taeda wood chips using ethanol-water mixtures and carbon dioxide at high pressuresThe Journal of Supercritical Fluids20053139

46. B. B. Hallac, et alEffect of Ethanol Organosolv Pretreatment on Enzymatic Hydrolysis of Buddleja davidii Stem BiomassIndustrial & Engineering Chemistry Research, 201014671472

47. 24892499Brandt, A., et al., Ionic liquid pretreatment of lignocellulosic biomass with ionic liquid-water mixtures. Green Chemistry, 2011. 13(9): p. 2489-2499

48.   C. N. Hamelinck, G. v. Hooijdonk, and A. P. C. Faaij, Ethanol from lignocellulosic biomass: techno-economic performance in short-, middle- and long-termBiomass and Bioenergy2005384410

49.   M. Sedlak, and N. Ho, Production of ethanol from cellulosic biomass hydrolysates using genetically engineered saccharomyces yeast capable of cofermenting glucose and xylose.Applied biochemistry and biotechnology2004403416

50.   D. Humbird, and A. Aden, Biochemical Production of Ethanol from Corn Stover: 2008 State of Technology Model.2009National Renewable Energy Laboratory (NREL). NREL/TP-51046214

51.   R. E. Cripps, et alMetabolic engineering of Geobacillus thermoglucosidasius for high yield ethanol productionMetabolic Engineering, 2009398408

52.   R. Bezerra, and A. Dias, Enzymatic kinetic of cellulose hydrolysis: Inhibition by ethanol and cellobiose.Applied Biochemistry and Biotechnology20054959

53.   L. R. Lynd, et alConsolidated bioprocessing of cellulosic biomass: an update.Current Opinion in Biotechnology2005577583

54.   W. Van Zyl, et alConsolidated Bioprocessing for Bioethanol Production Using Biofuels, L. Olsson, Editor. 2007Springer Berlin / Heidelberg. 205235

55.   M. Jin, et alConsolidated bioprocessing (CBP) performance of Clostridium phytofermentans on AFEX-treated corn stover for ethanol production.Biotechnology and Bioengineering201112901297

56.   D. G. Olson, et alRecent progress in consolidated bioprocessing. Current Opinion in Biotechnology

57.   S. Phillips, et alThermochemical Ethanol via Indirect Gasification and Mixed Alcohol Synthesis of Lignocellulosic Biomass2007National Renewable Energy Laboratory (NREL). NREL/TP-51041168

58.   A. Dutta, et alAn economic comparison of different fermentation configurations to convert corn stover to ethanol using Z. mobilis and Saccharomyces.Biotechnology Progress20106472

59.   A. L. Stephenson, et alThe environmental and economic sustainability of potential bioethanol from willow in the UKBioresource Technology201096129623

60. P. Sassner, M. Galbe, and G. Zacchi, Techno-economic evaluation of bioethanol production from three different lignocellulosic materialsBiomass and Bioenergy2008422430

61. L. Wang, et alTechnology performance and economic feasibility of bioethanol production from various waste papersEnergy & Environmental Science 201257175730

62. D. Mu, et alComparative Life Cycle Assessment of Lignocellulosic Ethanol Production: Biochemical Versus Thermochemical Conversion.Environmental Management2010565578

63. N. Qureshi, andT. C. Ezeji, Butanol, 'a superior biofuel' production from agricultural residues (renewable biomass): recent progress in technologyBiofuelsBioproducts and Biorefining, 2008319330

64. T. Ezeji, N. Qureshi, and H. P. Blaschek, Butanol production from agricultural residues: Impact of degradation products on Clostridium beijerinckii growth and butanol fermentation. Biotechnology and Bioengineering200714601469

65. R. Cascone, Biobutanol- a replacement for bioethanol? Chemical Engineering Progressing, 2008S4S9

66. P. H. Pfromm, et alBio-butanol vs. bio-ethanol: A technical and economic assessment for corn and switchgrass fermented by yeast or Clostridium acetobutylicumBiomass and Bioenergy, 2010515524

67. K. M. Shereena, and T. Thangaraj, Biodiesel: an alternative fuel produced from vegetable oils by transesterification. Electronic Journal of Biology, 20096774

68. M. Zabeti, W. M. A. Wan, Daud, and M.K. Aroua, Activity of solid catalysts for biodiesel production: A reviewFuel Processing Technology2009770777

69. F. Ma, and M. A. Hanna, Biodiesel production: a review. Bioresource Technology, 1999115

70. J. M. Marchetti, V. U. Miguel, and A. F. Errazu, Possible methods for biodiesel productionRenewable and Sustainable Energy Reviews200713001311

71. R. Schnepf, Agriculture-Based Biofuels: Overview and Emerging2010Congressional Research Service. R41282.

72. S. Tamalampudi, et alEnzymatic production of biodiesel from Jatropha oil: A comparative study of immobilized-whole cell

and commercial lipases as a biocatalystBiochemical Engineering Journal2008185189

73. H. Lu, et alProduction of biodiesel from Jatropha curcas L. oil. Computers & Chemical Engineering, 200910911096

74. S. Shah, and M. N. Gupta, Lipase catalyzed preparation of biodiesel from Jatropha oil in a solvent free systemProcess Biochemistry2007409414

75. A. Kumari, et alEnzymatic transesterification of Jatropha oilBiotechnology for Biofuels, 20091

76. Y. Chisti, Biodiesel from microalgae. Biotechnology Advances, 2007294306

77. S.A Scott, ., et al., Biodiesel from algae: challenges and prospects. Current Opinion in Biotechnology, 2010. 21(3): 277286 .

78. J. Grobbelaar, Turbulence in mass algal cultures and the role of light/dark fluctuationsJournal of Applied Phycology1994331335

79. Y. Chisti, Biodiesel from microalgae beats bioethanol.Trends in Biotechnology2008126131

80. K. D. Monson, et alAnaerobic digestion of biodegrabale municiple solid wastes: A review. 2007University of Glamorgan

81. A. P. Trzcinski, Anaerobic membrane bioreactor technology for solid waste stabilizationin Chemical Engineering and Chemical Technology. 2009Imperial College London: London.

82. J. Rapport, et alCurrent anaerobic digestion technologies used for treatment of municipal organic solid waste. 2008Department of Biological and Agricultural Engineering, University of California Davis, CA

83. F. Raposo, et alInfluence of inoculum to substrate ratio on the biochemical methane potential of maize in batch testsProcess Biochemistry200614441450

84. H. Hartmann, H. B. Moller, and B. K. Ahring, Efficiency of the anaerobic treatment of the organic fraction of municipal solid waste: collection and pretreatment.Waste Management & Research, 20043541

85. WRAPAnaerobic digestate. 2008Waste & Resources Action Programme (WRAP)

86. W. Edelmann, K. Schleiss, and A. Joss, Ecological, energetic and economic comparison of anaerobic digestion with different competing technologies to treat biogenic wastes.Water Science and Technology, 2000263273

87. W. Edelmann, U. Baier, and H. Engeli, Environmental aspects of the anaerobic digestion of the organic fraction of municipal solid wastes and of solid agricultural wastes.Water Science and Technology, 2004203208

88. D. R. Dodds, and R. A. Gross, Chemicals from Biomass. Science, 200712501251

89. Carlson, T.L. and E.M. Peters. 2006. Patent application US2006/094093 A1

90. P. Lorenz, and H. Zinke, White biotechnology: differences in US and EU approaches? Trends in Biotechnology, 2005570574

91. Y. K. Jung, et alMetabolic engineering of Escherichia coli for the production of polylactic acid and its copolymers.Biotechnology and Bioengineering2010161171

92. 19591966Wyman, C.E., et al., Coordinated development of leading biomass pretreatment technologies. Bioresource Technology, 2005. 96(18): p. 1959-1966

93. R. Yáñez, et alProduction of D(-)-lactic acid from cellulose by simultaneous saccharification and fermentation using &lt;i&gt;Lactobacillus coryniformis&lt;/i&gt; subsp. &lt;i&gt;torquens&lt;/i&gt;. Biotechnology Letters, 200311611164

94. M. G. Adsul, et alDevelopment of biocatalysts for production of commodity chemicals from lignocellulosic biomassBioresource Technology201143044312

95. H. Danner, et alBacillus stearothermophilus for thermophilic production of l-lactic acid.Applied Biochemistry and Biotechnology1998895903

96. I. S. Maddox, N. Qureshi, and K. Roberts-thomson, Production of acetone-butanol-ethanol from concentrated substrate using clostridium acetobutylicum in an integrated fermentation-product removal processProcess Biochemistry1995209215

97. D. T. Jones, and D. R. Woods, Acetone-butanol fermentation revisited. Microbiology and Moclecular Biology Reviews, 1986484524

98. A. Faik, Xylan biosysthesis: News from the grass.Plant Physiology2010396402

99. T. Granström, K. Izumori, and M. Leisola, A rare sugar xylitol. Part II: biotechnological production and future applications of xylitol.Applied Microbiology and Biotechnology2007273276

100. B. Saha, Hemicellulose bioconversion. Journal of Industrial MicrobiologyBiotechnology, 2003279291

101. M. Choct, and G. Annison, Anti-nutritive activity of wheat pentosans in broiler diets.British Poultry Science1990811821

102. S. J. B. Duff, and W. D. Murray, Bioconversion of forest products industry waste cellulosics to fuel ethanol: a reviewBioresource Technology1996133

103. D. J. Hayes, et alThe Biofine Process- Production of Levulinic Acid, Furfural, and Formic Acid from Lignocellulosic Feedstocks, in Biorefineries-Industrial Processes and Products. 2008Wiley-VCH Verlag GmbH. 139164

104. Fatih DemirbasM., Biorefineries for biofuel upgrading: A critical reviewApplied Energy2009Supplement 1(0): S151S161

105. J. N. Saddler, Bioconversion of Forest and Agricultural Plant ResiduesBiotechnology in agriculture. 1993CABI, Wallingford, UK.

106. Y. Sun, and J. Y. Cheng, Hydrolysis of lignocellulosic materials for ethanol production: a review.Bioresource Technology2002111

107. S. zhu, et alfed-batch simultaneous saccharification and fermentation of microwave/acid/alkali/$H_2O_2$ pretreated rice straw for production of ethanol.Chemical Engineering Communications2006639648

108. E. G. Koukios, and G. N. Valkanas, Process for chemical separation of the three main components of lignocellulosic biomassIndustrial & Engineering Chemistry Product Research and Development, 1982309314

109. E. G. Koukios, Biomass refining: a non-waste approach., in Economics and Ecosystem Management, D.O. Hall, N. Myers, and N.S. Margaris, Editors. 1985Dr W. Junk Publishers: Dordrecht, The Netherlands. 233244

110. B. Kamm, P. R. Gruber, and M. Kamm, Biorefineries- Industrial Processes and Products, in Ullmann's Encyclopedia of Industrial Chemistry. 2000Wiley-VCH Verlag GmbH & Co. KGaA.

111. T. N. Kleinert, Organosolv pulping with aqueous alcohol. Tappi Journal, 197499102

112. J. H. Lora, and S. Aziz, Organosolv pulping- a versatile approach to wood refining. Tappi Journal, 19859497

113. K. V. Sarkanen, Chemistry of solvent pulping. Tappi Journal, 1990215219

114. X. Pan, et alBioconversion of hybrid poplar to ethanol and co-products using an organosolv fractionation process: Optimization of process yields.Biotechnology and Bioengineering2006851861

115. X. Pan, et alEffect of organosolv ethanol pretreatment variables on physical characteristics of hybrid poplar substrates.Applied Biochemistry and Biotechnology2007367377

116. X. Pan, et alThe bioconversion of mountain pine beetle-killed lodgepole pine to fuel ethanol using the organosolv process. Biotechnology and Bioengineering20083948

117. M. J. Taherzadeh, and K. Karimi, Pretreatment of lignocellulosic wastes to improve ethanol and biogas production: A reviewInternationalJournalofMolecularSciences,200816211651

118. H. L. Chum, et alOrganosolv pretreatment for enzymatic hydrolysis of poplars: I. Enzyme hydrolysis of cellulosic residues. Biotechnology Bioengineering, 1988

119. X. Zhao, K. Cheng, and D. Liu, Organosolv pretreatment of lignocellulosic biomass for enzymatic hydrolysisApplied Microbiology and Biotechnology2009815827

120. A. Johansson, O. Aaltonen, and P. Ylinen, Organosolv pulping-methods and pulp properties. Biomass, 19874565

121. S. Aziz, and K. V. Sarkanen, Organosolv pulping-a review. Tappi Journal, 1989169175

122. R. W. Thring, E. Chornet, and R. P. Overend, Recovery of a solvolytic lignin: Effects of spent liquor/acid volume ratio, acid concentration and temperature.Biomass1990289305

123. D. Sidiras, and E. Koukios, Simulation of acid-catalysed organosolv fractionation of wheat straw.Bioresource Technology20049198

124. J. Y. Zhu, and X. J. Pan, Woody biomass pretreatment for cellulosic ethanol production: Technology and energy consumption evaluationBioresource Technology201049925002

125. Myllymaki, V. and R. Aksela. 2005. Dissolution and delignification of lignocellulosic materials with ionic liquid solvent under microwave irradiation. WO patent 2005/017001

126. D. J. Hayes, An examination of biorefining processes, catalysts and challengesCatalysis Today2009138151

127. A. P. Dadi, S. Varanasi, and C. A. Schall, Enhancement of cellulose saccharification kinetics using an ionic liquid pretreatment step. Biotechnology and Bioengineering2006904910

128. S. Zhu, Use of ionic liquids for the efficient utilization of lignocellulosic materialsJournal of Chemical Technology & Biotechnology, 2008777779

129. W. M. Reichert, et alDerivatization of chitin in room temperature ionic liquids, in 222 ACS National Meeting. 2001Chicago, IL, United States.

130. T. Heinze, K. Schwikal, and S. Barthel, Ionic Liquids as Reaction Medium in Cellulose Functionalization.Macromolecular Bioscience2005520525

131. B. S. Dien, et alEnzymatic saccharification of hot-water pretreated corn fiber for production of monosaccharidesEnzyme and Microbial Technology200611371144

132. M. Negro, et alHydrothermal pretreatment conditions to enhance ethanol production from poplar biomassApplied Biochemistry and Biotechnology200387100

133. T. Rogalinski, T. Ingram, and G. Brunner, Hydrolysis of lignocellulosic biomass in water under elevated temperatures and pressuresThe Journal of Supercritical Fluids20085463

134. G. Van Walsum, et alConversion of lignocellulosics pretreated with liquid hot water to ethanolApplied Biochemistry and Biotechnology, 1996157170

135. E. U. Franck, Fluids at high pressures and temperatures. Chemical thermodynamics, 1987225242

136. R. W. Shaw, et alSupercritical water. A medium for chemistry. Chemical Engineering News, 19912639

137. P. E. Savage, et alReactions at supercritical conditions: {A}pplications and fundamentals. AIChE Journal, 199517231778

138. A. R. Katritzky, et alReactions in High-Temperature Aqueous MediaChemical Reviews2001837892

139. K. L. Kohlmann, et alEnhanced Enzyme Activities on Hydrated Lignocellulosic Substratesin Enzymatic Degradation of Insoluble Carbohydrates. 1996American Chemical Society. 237255

140. J. Weil, et alContinuous pH monitoring during pretreatment of yellow poplar wood sawdust by pressure cooking in waterApplied Biochemistry and Biotechnology199899111

141. P. Alvira, et alPretreatment technologies for an efficient bioethanol production process based on enzymatic hydrolysis: A review. Bioresource Technology201048514861

142. D. A. I. Goring, Thermal softening of lignin, hemicelluloses and cellulose. 1963Pulp and Paper Research Institute of Canada.

143. R. P. Overend, E. Chornet, and J. A. Gascoigne, Fractionation of Lignocellulosics by Steam-Aqueous Pretreatments [and Discussion]Philosophical Transactions of the Royal Society of London. Series A, Mathematical and Physical Sciences, 1987523536

144. W. H. Mason, 1929Apparatus and process of explosion defibration of lignocellulosic material. U.S. Patent 1655618

145. D. Montane, et alApplication of steam explosion to the fractionation and rapid vapor-phase alkaline pulping of wheat strawBiomass and Bioenergy1998261276

146. X. Pan, et alStrategies to enhance the enzymatic hydrolysis of pretreated softwood with high residual lignin content.Applied Biochemistry and Biotechnology200510691079

147. R. Eklund, M. Galbe, and G. Zacchi, The influence of $SO_2$ and $H_2SO_4$ impregnation of willow prior to steam pretreatmentBioresource Technology1995225229

148. J. E. Fein, D. Potts, and D. Good, Development of an optimal wood-to-fuel ethanol process utilizing best available technology. Energy Biomass and Waste, 1991

149. M. Ropars, et alLarge-scale enzymatic hydrolysis of agricultural lignocellulosic biomass. Part 1: Pretreatment proceduresBioresource Technology1992197204

150. P. J. Morjanoff, and P. P. Gray, Optimization of steam explosion as method for increasing susceptibility of sugarcane bagasse to enzymatic saccharification. Biotechnology Bioengineering, 1987733741

151. B Focher, ., A Marzetti, , and V Crescenzi, , Steam Explosion Techniques, Fundamentals and Industrial Applications. 1991, Gordon and Breach Publishers: Philadelphia. 412

152. B. K. Avellar, and W. G. Glasser, Steam-assisted biomass fractionation. I. Process considerations and economic evaluationBiomass and Bioenergy1998205218

153. Y. Zheng, and G. T. Tsao, Avicel hydrolysis by cellulase enzyme in supercritical CO2Biotechnology Letters1996451454

154. Y. Zheng, et alSupercritical carbon dioxide explosion as a pretreatment for cellulose hydrolysisBiotechnology Letters1995845850

155. C. Schacht, C. Zetzl, and G. Brunner, From plant materials to ethanol by means of supercritical fluid technologyThe Journal of Supercritical Fluids2008299321

156. L Laureano-Perez, ., et al., Understanding Factors that Limit Enzymatic Hydrolysis of Biomass

157. nty-Sixth Symposium on Biotechnology for Fuels and ChemicalsB.H. Davison, et al., Editors. 2005Humana Press. 10811099

158. 20142018Teymouri, F., et al., Optimization of the ammonia fiber explosion (AFEX) treatment parameters for enzymatic hydrolysis of corn stover. Bioresource Technology, 2005. 96(18): p. 2014-2018

159. 20192025Eggeman, T. and R.T. Elander, Process and economic analysis of pretreatment technologies. Bioresource Technology, 2005. 96(18): p. 2019-2025

160. S. Ghosh, et alPilot-scale gasification of municipal solid wastes by high-rate and two-phase anaerobic digestion (TPAD).Water Science and Technology 2000101110

161. M. Beccari, et alEnhancement of anaerobic treatability of olive oil mill effluents by addition of $Ca(OH)_2$ and bentonite without intermediate solid/liquid separation. Water Science and Technology, 2001275282

162. S. Tanaka, et alEffects of thermochemical pretreatment on the anaerobic digestion of waste activated sludgeWater Science and Technology, 1997209215

163. H. H. Brownell, E. K. C. Yu, and J. N. Saddler, Steam-explosion pretreatment of wood: Effect of chip size, acid, moisture content and pressure dropBiotechnology and Bioengineering, 1986792801

164. D Gregg, and J Saddler,, A techno-economic, assessment of the pretreatment and fractionation steps of a biomass-to-ethanol process. Applied Biochemistry and Biotechnology, 1996. 57-58(1): 711727 .

165. V. S. Chang, et alSimultaneous saccharification and fermentation of lime-treated biomassBiotechnology Letters200113271333

166. X. F. Sun, et alCharacteristics of degraded cellulose obtained from steam-exploded wheat strawCarbohydrate Research200597106

167. J. L. Jones, and K. T. Semrau, Wood hydrolysis for ethanol production- previous experience and the economics of selected processes. Biomass, 1984109135

168. A. M. J. Kootstra, et alComparison of dilute mineral and organic acid pretreatment for enzymatic hydrolysis of wheat strawBiochemical Engineering Journal2009126131

169. A. B. Bjerre, et alPretreatment of wheat straw using combined wet oxidation and alkaline hydrolysis resulting in convertible cellulose and hemicelluloseBiotechnology and Bioengineering1996568577

170. P. Kumar, et alMethods for Pretreatment of Lignocellulosic Biomass for Efficient Hydrolysis and Biofuel ProductionIndustrial & Engineering Chemistry Research, 200937133729

171. F. Carvalheiro, L. C. Duarte, and F. M. Gírio, Hemicellulose biorefineries: a review on biomass pretreatmentsScientific & Industrial Research, 2008849864

172. Molero GómezA., C. Pereyra López, and E. Martinez de la Ossa, Recovery of grape seed oil by liquid and supercritical carbon dioxide extraction: a comparison with conventional solvent extraction. The Chemical Engineering Journal and the Biochemical Engineering Journal, 1996227231

173. K. L. Toews, et alpH-Defining Equilibrium between Water and Supercritical $CO_2$. Influence on SFE of Organics and Metal ChelatesAnalytical Chemistry199540404043

174. P. Intanakul, M. Krairiksh, and P. Kitchaiya, Enhancement of Enzymatic Hydrolysis of Lignocellulosic Wastes by Microwave Pretreatment Under Atmospheric PressureJournal of Wood Chemistry and Technology2003217225

175. B. C. Saha, A. Biswas, and M. A. Cotta, Microwave Pretreatment, Enzymatic Saccharification and Fermentation of Wheat Straw to EthanolJournal of Biobased Materials and Bioenergy2008210217

176. H. Ma, et alEnhanced enzymatic saccharification of rice straw by microwave pretreatmentBioresource Technology200912791284

177. M. Imai, K. Ikari, and I. Suzuki, High-performance hydrolysis of cellulose using mixed cellulase species and ultrasonication pretreatmentBiochemical Engineering Journal20047983

178. S. Nitayavardhana, et alUltrasound pretreatment of cassava chip slurry to enhance sugar release for subsequent ethanol production.Biotechnology and Bioengineering2008487496

179. D. R. Keshwani, Microwave Pretreatment of Switchgrass for Bioethanol Production2009North Carolina State University.

180. R. C. Sun, and J. Tomkinson, Characterization of hemicelluloses obtained by classical and ultrasonically assisted extractions from wheat strawCarbohydrate Polymers2002263271

181. V. Yachmenev, et alAcceleration of the enzymatic hydrolysis of corn stover and sugar cane bagasse celluloses by low intensity uniform ultrasoundBiobased Material Bioenergy, 20092531

182. M. e. a. Vinatoru, Ultrasonically assisted extraction of bioactive principles from plants and their constituents, in Advances in Sonochemistry, T.J. Mason, Editor. 1999JAI Press. 209248

183. H. Palonen, et alEvaluation of wet oxidation pretreatment for enzymatic hydrolysis of softwoodApplied Biochemistry and Biotechnology, 2004117

184. A. S. Schmidt, and A. B. Thomsen, Optimization of wet oxidation pretreatment of wheat strawBioresource Technology1998139151

185. H. L. Chum, et alEvaluation of pretreatments of biomass for enzymatic hydrolysis of cellulose1985Solar Energy Research Institute: Golden, Colorado. 64

186. C. Martín, et alWet oxidation pretreatment, enzymatic hydrolysis and simultaneous saccharification and fermentation of clover-ryegrass mixturesBioresource Technology200887778782

187. M. Galbe, and G. Zacchi, A review of the production of ethanol from softwoodAppl Microbiol Biotechnol, 2002618628

188. P. Vidal, Ozonolysis of Lignin- Improvement of in vitro digestibility of poplar sawdust. Biomass, 1988117

189. Energy Information Administration, Chapter 10 Growth of the Landfill Gas Industry, in Renewable Energy Annual 1996. 1997. DOE/EIA-0603(96

190. A. Breen, and F. L. Singleton, Fungi in lignocellulose breakdown and biopulpingCurrent Opinion in Biotechnology1999252258

191. R. Hischier, Life Cycle Inventories of Packaging and Graphical Papers, in Ecoinvent Report 112007Swiss Centre for Life Cycle Inventories: Dübendorf

192. Á. T. Martínez, et alBiodegradation of lignocellulosics: microbial, chemical, and enzymatic aspects of the fungal attack of lignin. International micorbiology, 2005195204

193. P. Maijala, et alBiomechanical pulping of softwood with enzymes and white-rot fungus Physisporinus rivulosusEnzyme and Microbial Technology2008169177

194. Shukla, O.P., U.N. Rai, and S.V. Subramanyam. Biopulping and Biobleaching: An Energy and envioronment Saving Technology for Indian Pulp and Paper Industry. 2004 [cited 2012 April]; Available from: http://isebindia.com/01_04/04-04-3.html

195. M. J. Ray, et alBrown rot fungal early stage decay mechanism as a biological pretreatment for softwood biomass in biofuel productionBiomass and Bioenergy, 201012571262

196. C. Wan, and Y. Li, Fungal pretreatment of lignocellulosic biomassBiotechnology Advances

197. H. Yu, et alThe effect of biological pretreatment with the selective white-rot fungus Echinodontium taxodii on enzymatic hydrolysis of softwoods and hardwoodsBioresource Technology200951705175

198. M. Monrroy, et alStructural change in wood by brown rot fungi and effect on enzymatic hydrolysisEnzyme and Microbial Technology2011472477

# Citations

## CHAPTER 1

Yangping Zhou, Fu Li, Zhiwei Zhou, and Yuanle Ma, "Thermal Hydraulic Analysis Using GIS on Application of HTR to Thermal Recovery of Heavy Oil Reservoirs," Science and Technology of Nuclear Installations, vol. 2012, Article ID 676529, 15 pages, 2012. doi:10.1155/2012/676529.

## CHAPTER 2

an Dolfing, Stephen R Larter, and Ian M Head, Thermodynamic Constraints on Methanogenic Crude Oil Biodegradation, doi:10.1038/is-mej.2007.111.

# CHAPTER 3

Head IM, Gray ND and Larter SR (2014) LJife in the slow lane; biogeo-chemistry of biodegraded petroleum containing reservoirs and implications for energy recovery and carbon management. Front. Microbiol. 5:566. doi: 10.3389/fmicb.2014.00566.

# CHAPTER 4

Moon Sik Jeong, Jinhyung Cho, Jinsuk Choi, Ji Ho Lee, and Kun Sang Lee, "Compositional Simulation on the Flow of Polymeric Solution Alternating $CO_2$ through Heavy Oil Reservoir,"Advances in Mechanical Engineering, vol. 2014, Article ID 978465, 9 pages, 2014, doi:10.1155/2014/978465.

# CHAPTER 5

You-Hong Sun, Feng-Tian Bai, Xiao-Shu Lü, Qiang Li, Yu-Min Liu, Ming-Yi Guo, Wei Guo, and Bao-Chang Liu, A Novel Energy-efficient Pyrolysis Process: Self-pyrolysis of Oil Shale Triggered by Topochemical Heat in a Horizontal Fixed Bed, doi:10.1038/srep08290.

# CHAPTER 6

N. Abdullah and F. Sulaiman (2013). The Oil Palm Wastes in Malaysia, Biomass Now - Sustainable Growth and Use, Dr. Miodrag Darko Matovic (Ed.), ISBN: 978-953-51-1105-4, IJnTech, DOI: 10.5772/55302.

# CHAPTER 7

Hongbin Cheng and Lei Wang (2013). Lignocelluloses Feedstock Biorefinery as Petrorefinery Substitutes, Biomass Now - Sustainable Growth and Use, Dr. Miodrag Darko Matovic (Ed.), ISBN: 978-953-51-1105-4, InTech, DOI: 10.5772/51491.

# Index